Laboratory Guide to Enzymology

Laboratory Guide to Enzymology

Geoffrey A. Holdgate, Antonia Turberville, and Alice Lanne

Library of Congress Cataloging-in-Publication Data

Names: Holdgate, Geoffrey A., author. | Turberville, Antonia, author. | Lanne, Alice, author.
Title: Laboratory guide to enzymology / Geoffrey A. Holdgate, Antonia Turberville, Alice Lanne.
Description: Hoboken, New Jersey : Wiley, [2024] | Includes bibliographical references and index.
Identifiers: LCCN 2023046724 (print) | LCCN 2023046725 (ebook) | ISBN 9781394179794 (paperback) | ISBN 9781394179800 (adobe pdf) | ISBN 9781394179824 (epub)
Subjects: MESH: Enzymes–pharmacokinetics | Enzyme Assays–methods | Drug Discovery
Classification: LCC QP601 (print) | LCC QP601 (ebook) | NLM QU 135 | DDC 572/.7–dc23/eng/20231128
LC record available at https://lccn.loc.gov/2023046724
LC ebook record available at https://lccn.loc.gov/2023046725

Cover Design: Wiley
Cover Images: Courtesy of Geoffrey A. Holdgate; © StudioMolekuul/Shutterstock

Set in 9.5/12.5pt STIXTwoText by Straive, Chennai, India

Contents

Preface

The application of enzymology is an essential approach to drug discovery. Many drug targets are enzymes and modulating their behavior can provide useful therapeutic intervention. As such, an understanding of the basic principles in the use of enzymes is important in identifying and characterizing molecules that may change the function of enzymes. Sometimes, however, the topic can appear difficult as it combines chemical and mathematical concepts that are often unfamiliar to the biologist. Fortunately, there are many useful textbooks that provide information in great detail, covering the many aspects of enzymology. This book is not meant to be a replacement for those textbooks, nor is it a repository for information; rather, it is a guide that helps readers navigate the field and gain a basic understanding of the principles and techniques used in the study of enzymes. In writing this book, we focused on providing the fundamental principles, concepts, protocols, and examples required to generate and analyze enzyme kinetic data. The book serves to provide an initial text that allows the reader to undertake and understand the experiments that are required in establishing assays and building screening approaches that are the bedrock of early drug discovery.

The book begins with an introduction to proteins and enzymes, illustrating the structural features that are key to protein function. Aspects of binding kinetics and thermodynamics are introduced and the importance of quality control when working with proteins is highlighted. The use of buffers in enzyme studies is described as the control of pH is a critical requirement when working to characterize enzymes and their inhibitors.

Steady-state assays and their design to understand enzyme function and to identify and characterize inhibitors and activators are discussed. The different types of mechanism of action for these modulators with respect to substrate concentration are introduced and data analysis methods are presented.

Finally, the different types of molecular interactions are presented and key examples of application in drug discovery are described. The appendices contain a range of key information that supplements the material in the main text.

We hope that this book will be a useful laboratory companion for life science students, academic and industrial researchers who are interested in learning about enzymology. It is our goal to provide a clear and concise introduction to the field, taking a stepwise journey through the equations and their derivations so that the mathematics is not overwhelming and the rationale for the models used is clear. In this way, we trust that the book will inspire readers to begin their journey into exploring enzymes, their mechanisms, and regulation.

Cheshire, 2024

Geoffrey A. Holdgate
Antonia Turberville
Alice Lanne

Acknowledgments

We are grateful to many people over many years for their contributions in turning the idea of a laboratory guide to enzymology into reality. This project certainly would not have been possible without the guidance and mentorship of Wal Ward. His experience and expertise guided enzyme kinetic studies at AstraZeneca and its predecessor companies for two decades. Without his teaching of the fundamental principles, his drive for quality, and wise advice, we would not have been able to continue to deliver leadership in enzymology to the company over the last 10 years. During that time, we have also been indebted to numerous other current and previous colleagues, including Rachel Grimley, Archie Argyrou, Bharath Srinivasan, Xiang Zhai, Astrid Kraal, Kara Herlihy, Hua Xu, Bryony Ackroyd, Sheffin Joseph, and Christopher Stubbs who have contributed ideas, discussion, challenge, and support. We especially thank Bharath Srinivasan and Gareth Davies for critical reading of the manuscript.

1

Introduction to Proteins and Enzymes

CHAPTER MENU

1.1 Protein Structure

Proteins are the central functional molecules of life, encoded by DNA, translated, and expressed to carry out the essential functions of the cell. The building blocks for proteins are amino acids: every amino acid contains a positively charged amine group (N-terminus), a negatively charged carboxyl group (C-terminus), a hydrogen atom, and an R group, all centered around a chiral carbon (alpha carbon, C_α) (Figure 1.1). The presence of a chiral carbon results in stereoisomerism; naturally occurring amino acids are *L*-isomers, and *D*-isomers can arise during chemical synthesis. There are 20 different R groups, which give rise to 20 different amino acids (Figure 1.2). Amino acids can be charged (negatively and positively), polar and non-polar. These different properties contribute to different bonding interactions and architecture of the protein (Section 1.1.4) [1, 2].

1.1.1 Primary Structure

Each protein is formed of a unique sequence of amino acids, which determines the properties of the protein. These are linked by covalent peptide bonds between the amino group of one residue and the carboxyl group of the next, forming long polypeptide chains of amino acids. The number and sequence of amino acids in a polypeptide chain is known as the primary (1°) structure of a protein and is determined by the DNA sequence of the gene. Mutations to the DNA sequence may lead to changes in the amino acids in the polypeptide chain, thus altering the primary structure of the protein [1, 2].

1.1.2 Secondary Structure

The secondary structure of proteins describes the layout of the protein backbone in three dimensions. This structure is formed from the individual peptide bonds between residues, which usually are planar and *trans* (with the exception of proline). There are common elements that often combine to contribute to the protein backbone describing its overall fold. Rotations around the

Laboratory Guide to Enzymology, First Edition. Geoffrey A. Holdgate, Antonia Turberville, and Alice Lanne.
© 2024 John Wiley & Sons, Inc. Published 2024 by John Wiley & Sons, Inc.

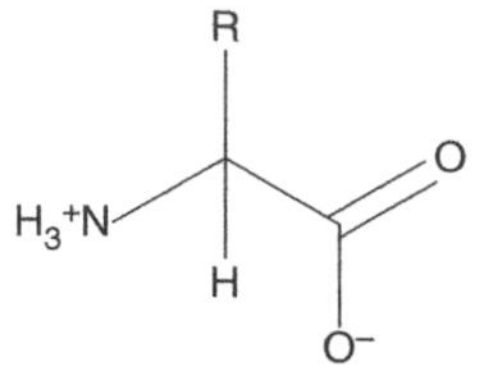

Figure 1.1   General amino acid structure.

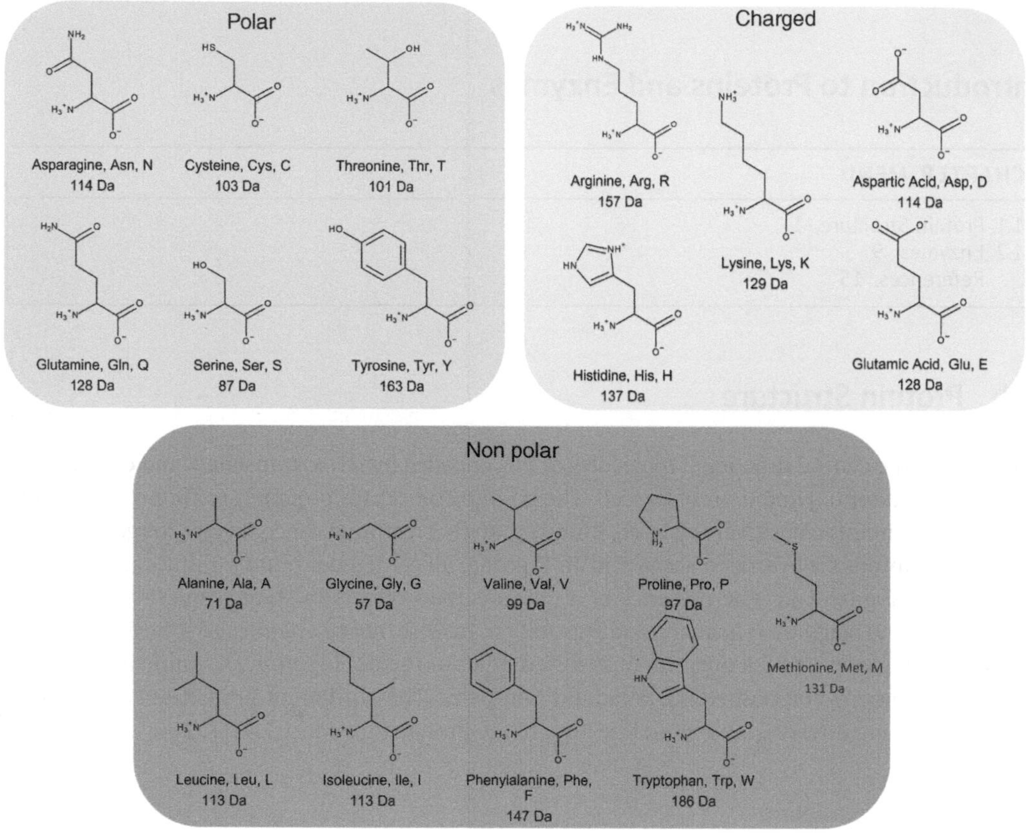

Figure 1.2   Chemical structure of amino acids.

peptide bond enable hydrogen bond formation between the carbonyl oxygen group and amide hydrogen atom of spatially adjacent amino acids, resulting in folding of the polypeptide chains into secondary (2°) structures. Hydrogen bonding can also occur between amino acid side chains. Common secondary structures include the alpha helix, the beta sheet, loops, and many protein structures contain a combination of all elements [1].

1.1.2.1   The Alpha Helix

One of the most important structural features is the alpha helix (Figure 1.3). This is a right-handed helical structure containing 3.6 amino acid residues in each turn. It is formed when each N-H group donates a hydrogen bond to the backbone C=O group of the amino acid four residues before it in the polypeptide chain. This occurs as the C=O groups in the helix are parallel to the axis and are directionally aligned with the N-H groups to which the hydrogen bond is formed. The amino acid side chains are positioned away from the axis. Alpha helices can vary in length, although there are few

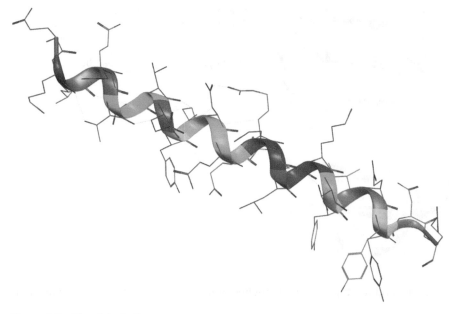

Figure 1.3 The alpha helix.
The structure of the alpha helix is shown: the backbone of the helix is represented in cartoon, and sticks show the amino acid side chains protruding from the backbone. The colors used are from the Clustal-X color scheme (Table 1.1).

examples of proteins where the helix length extends beyond 40 residues. Clearly, the first and last residue of an alpha helix cannot make hydrogen bonds to contribute to the helix, so these residues are often amino acids that can make hydrogen bonds with other parts of the protein or with the solvent. Some residues are more likely to form alpha helices than others, with alanine, leucine, arginine, methionine, and lysine having the highest propensity, although the tendency to form helices will depend on the identities of the neighboring residues. Conversely, residues such as aspartate, glycine, and proline tend not to form alpha helices. Proline cannot donate an amide hydrogen bond and also interferes sterically with the backbone of the preceding turn. However, proline may sometimes be positioned as the first residue in an alpha helix, providing structural rigidity to the helix. Often, alpha helices display an amphipathic nature, with hydrophobic residues located on one side and hydrophilic residues on the other. Another feature of alpha helices is that they tend to have a macrodipole, with the Nterminus being the positive pole. This arises as the individual microdipoles from the carbonyl groups of the peptide bonds in the helix align along the axis [1, 2].

1.1.2.2 The Beta Sheet

Another common structural motif in proteins is the beta sheet (Figure 1.4). When the backbone of a protein exists in an extended conformation (beta strand), it is possible for residues to make complementary hydrogen bonds with another beta strand. These interactions may occur when the chains are aligned in the same or opposite directions. When the chains are aligned in the same direction, the arrangement is termed a parallel beta sheet, and when the chains alternate in direction, it is termed an antiparallel beta sheet. Usually, an extensive hydrogen bond network is established where the N—H groups in the backbone of one strand establish hydrogen bonds with the C=O groups in the backbone of the adjacent strand. Often, beta sheets contain around 10 residues but can be much shorter (as low as 2 or 3 residues). Beta sheets often contain large aromatic residues

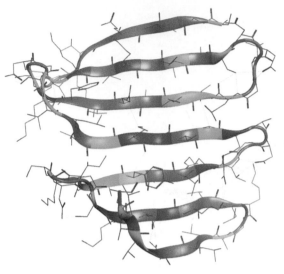

Figure 1.4 **The beta sheet.**
The structure of a beta sheet is shown: the backbone of the sheet is represented in cartoon, and sticks show the amino acid side chains protruding from the backbone.

(tyrosine, phenylalanine, and tryptophan) and branched amino acids (threonine, isoleucine, and valine) [1, 2].

1.1.2.3 Loops

There are segments of a protein that connect the alpha helix and beta sheet elements together, which in themselves do not have recognizable regular structural patterns. These secondary structural elements are termed loops (Figure 1.5). Loops are an important component of secondary structure, often containing as much as half of the total number of residues in a protein [3]. Loops often contribute significantly to the overall shape, dynamics, and physicochemical properties of the protein [4]. Loops are frequently located on the protein's surface in solvent-exposed

Figure 1.5 **Loop region.**
The structure of a loop region is shown: the backbone of the sheet is represented in cartoon, and sticks show the amino acid side chains protruding from the backbone.

Table 1.1 Clustal-X color scheme for coloring amino acids.

Clustal-X color scheme		
Category	**Color**	**Residue**
Hydrophobic	Blue	A, I, L, M, F, W, V
Positive charge	Red	K, R
Negative charge	Magenta	E, D
Polar	Green	N, Q, S, T
Cysteine	Pink	C
Glycine	Orange	G
Proline	Yellow	P
Aromatic	Cyan	H, Y
Unconserved	White	Any, gap

regions and are often involved in important interactions. Despite the lack of patterns, loops do not appear to be completely random structures, and they have been classified in various ways, including their geometrical shape [5]. However, even though their importance is recognized, loop structure remains difficult to predict.

The primary structure of a protein influences the secondary structure, with certain residues more likely to form one structure over the other; for example, proline residues are often called "helix breakers" as their cyclic nature induces a kink in the polypeptide chain and prevent alpha helix formation. Glycine residues, for example, also are frequently involved in tight turns as they are small and flexible [2].

The image for the loop structure has been colored by structure (in the program MOE2022; red: alpha helix, yellow: beta sheet, loop: white, turn: blue). The alpha helix and beta sheet above (Figures 1.3 and 1.4) have been colored using the Clustal-X color scheme (Table 1.1).

1.1.3 Tertiary Structure

The three-dimensional (3D) structure of a protein is defined by the position of all the atoms of the polypeptide chain arranged in 3D space. This is termed the tertiary (3°) structure, and it comprises the arrangement of the secondary structural elements, as described in Section 1.1.2, and involves numerous interactions between residues (Table 1.2). Proteins often contain regions that are conserved across or within families and which carry out similar functions in each. These are distinct folded units in a polypeptide chain that provide structural or functional features. For example, a protein could contain regulatory and catalytic domains. These structural elements may be domains, folds, and motifs. The combination of these elements in a single polypeptide chain may be sufficient to produce a fully functional protein without requirement for additional polypeptides [6].

1.1.3.1 Domains, Folds, and Motifs

Protein domains are areas of a protein that can fold stably and serve a specific purpose. Proteins may contain one or more domains (Figure 1.7). Each protein domain has a specific fold, which describes how the secondary structural elements in that domain are organized. Often, one fold

Table 1.2 Bonds and interactions in proteins.

Interactions	Description	Residues involved
Hydrogen bonds Interpeptide bond $$-C\!\!=\!\!\overset{\delta^-}{O}\!:\!-\!-\!-\!\overset{\delta^+}{H}\!-\!N\!-$$ Side chain bond (phenyl ring)$-O-\overset{\delta^+}{H}\!-\!-\!-\!:\!\overset{\delta^-}{O}\!\!=\!\!\overset{\overset{O^-}{\mid}}{C}\!-$	A hydrogen bond is the electrostatic interaction between a hydrogen atom (covalently bound to a residue) and the electronegative atom (O, N) of another residue. In proteins, this occurs between the hydrogen atom of the N–H group of one residue and the oxygen atom of the O=C of another residue.	Polar
Ionic bonds $$-H-\overset{\overset{H}{\mid}}{\underset{\underset{H}{\mid}}{N^+}}\!-\!-\!-\,O^-\!-\!\overset{\overset{}{\underset{\underset{O}{\parallel}}{C}}}\!-$$	An ionic bond is the electrostatic interaction between two groups of opposite charges.	Charged
Hydrophobic interactions $-R \quad R-$	Hydrophobic interactions describe the tendency of non-polar molecules to associate in aqueous solution, resulting largely from the favorable entropy produced by the breaking of the hydrogen bonds of the surrounding water.	Non-polar
Disulfide bonds $-S-S-$	Disulfide bonds are covalent interactions between two sulfur molecules; this type of interaction only occurs between cysteine side chains.	Cys

may be used by several different proteins to fulfill a range of activities. A structural motif is a small region of 3D structure that arises in a range of diverse proteins that may have a unique function. Some common protein motifs are shown in Figure 1.6.

An example of protein domains is shown in Figure 1.7.

Proteins are usually considered as layers of the secondary structural elements (alpha helices, α and beta sheets, β), and four types of proteins have been described based on the combinations of these elements. These are α/α (consisting of all α), β/β (consisting of all β), α/β (consisting of both α and β in varied regions), and α+β (where the α and β elements occur in different regions to each other).

1.1.4 Quaternary Structure

Whilst many proteins are made up of a single polypeptide chain, some proteins require multiple subunits to come together for functional activity. The coming together of multiple subunits forms the quaternary (4°) structure of a protein, and these can either be formed of multiple copies of the same (homopolymer) or different (heteropolymer) subunits. An example of a protein with a quaternary structure is hemoglobin, which is a heteropolymer formed of four subunits (two alpha and two beta subunits) [1, 2].

1.1.5 Protein Structure Prediction

The application of artificial intelligence has revolutionized protein structure prediction [7]. AlphaFold2 [8], DeepMind's machine-learning protein structure prediction program, released in

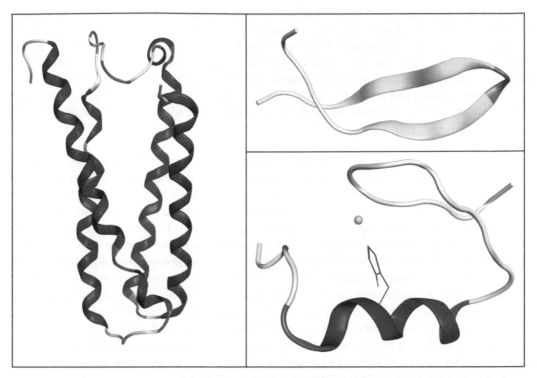

Figure 1.6 Common protein motifs.
Common protein motifs are shown in cartoon format: a four-helix bundle motif (left) is four α-helices packed together, typically in a lengthwise manner; a hairpin (top right) is a simple structure consisting of two antiparallel β-sheets joined by a loop; and a zinc finger motif (bottom right) is two beta strands with an alpha helix folded over to bind a zinc ion.

Figure 1.7 Multi-domain protein.
Pyruvate kinase contains an all-β-nucleotide-binding domain (in blue), an α/β-substrate-binding domain (in magenta), and an α/β-regulatory domain (in pink).

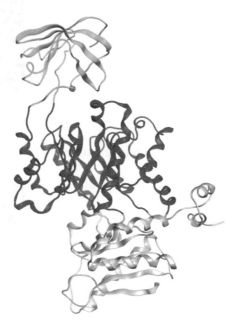

2021, enables the generation of high-confidence protein structures. This was the first indication that deep learning-based methods can now predict protein structures with an accuracy often comparable to that of experimental structures. AlphaFold2 generates 3D structures from input amino acid sequences by querying several databases of protein sequences and constructing a multiple sequence alignment (MSA). This enables the determination of the parts of the sequence that are more variable and allows the detection of correlations between them. The network also tries to identify proteins that may have a similar structure to the input, termed "templates", and constructs an initial "pair representation", which suggests which amino acids are likely to be in contact with each other. The MSA and the templates are passed through a transformer, which identifies the most useful information, and cyclically exchanges between the model and sequence alignment. Finally, the structure module incorporates the MSA representation and pair representation to construct a 3D model of the structure, which comprises a list of Cartesian coordinates representing the position of each atom of the protein. After generating a final structure, it is passed back through the process to refine the prediction. In 2022, AlphaFold2 was used to predict more than 200 million proteins from around 1 million different species. Now, another neural network, ESMFold [9] from Meta has been used to predict over 600 million proteins. Although this network is not currently as accurate as AlphaFold2, it is around 60 times faster at predicting structures for short sequences. Additionally, many of these structures are unlike anything in the current databases of experimentally determined protein structures or any of AlphaFold's predictions from known organisms. These breakthrough in artificial intelligence-based approaches demonstrate that structure prediction will become simpler, cheaper, and more accurate in future and will allow researchers to obtain structural information about proteins much more rapidly, allowing significant advances in the understanding of many areas of biology.

The AlphaFold source code can be found here.

https://github.com/deepmind/alphafold

The ESMFold source code can be found here.

https://github.com/facebookresearch/esm

1.2 Enzymes

Enzymes are proteins that carry out highly specific activities. Every enzyme contains an active site in which a specific molecule or reactant known as a substrate will bind, and the enzyme will convert the substrate into a product through a series of steps (Figure 1.4). Compounds that mimic the substrate are frequently used in drug discovery as a strategy to inhibit the targeted enzyme [10, 11].

The International Union of Pure and Applied Chemistry (IUPAC) Enzyme Commission (EC) developed a system to classify enzymes, to avoid confusion between different names for the same enzyme. This system is based on a numerical nomenclature, which describes the reaction catalyzed by a particular enzyme. Enzymes are split into six broad classes (Table 1.3), these are oxidoreductases, transferases, hydrolases, lyases, isomerases, and ligases [12]. Within these enzyme classes, there are further subdivisions of enzyme types. An example of a large subclass of enzymes are kinases, which phosphorylate their protein substrate (often another kinase) using adenosine triphosphate (ATP). This phosphorylation can act like a switch, activating or inactivating the protein substrate. Kinases are normally involved in complex signaling cascades in which there is a chain of kinases, each acting on the subsequent kinase in the cascade to bring about a biological effect.

Table 1.3 Enzyme classification.

Number	Class	Reaction type	Selected sub-classes
1	Oxidoreductases	Oxidation/reduction	1.1 acts on CH–OH group 1.2 acts on aldehyde group 1.3 acts on CH–CH group 1.4 acts on CH–NH_2 group
2	Transferases	Chemical group transfer reactions	2.1 transfers 1 carbon group 2.3 acyltransferases 2.4 glycosyltransferases 2.7 phosphotransferases
3	Hydrolases	Hydrolytic bond cleavage reactions	3.1 esterases 3.2 glycosidases 3.4 peptidases
4	Lyases	Non-hydrolytic bond cleavage or elimination reactions	4.1 C–C lyases 4.2 C–O lyases 4.3 C–N lyases
5	Isomerases	Rearrangement of atoms in molecules (isomerization)	5.1 racemases 5.3 intramolecular oxidoreductases 5.4 intramolecular transferases
6	Ligases	Bond synthesis to join two or more molecules together, coupled to hydrolysis (e.g. ATP)	6.1 C–O ligases 6.2 C–S ligases 6.3 C–N ligases

Notes: Enzymes in each class are subdivided using a second number, which more specifically defines the catalyzed reaction and classifies the sub-class. Third and fourth numbers classify each enzyme further into sub-sub-class and serial number to give each enzyme a unique identifier of the form EC 1.2.3.4, respectively.

1.2.1 Properties of Enzymes

1.2.1.1 Catalysis

Enzymes are biological catalysts. Like all catalysts, they increase the rate of a reaction without perturbing the equilibrium position (bringing about the same rate enhancement in either direction) and remain unchanged after the reaction. This enables the essential reactions carried out within the cell to proceed rapidly enough for metabolism to be maintained. Enzyme catalysis is usually discussed in terms of a model known as transition state theory. Enzymes increase the rate of reaction by reducing the free energy of the transition state (Figure 1.3). The transition state is defined as the most unstable species on the reaction pathway, and so occurs at a peak in the free energy profile of a reaction. Compounds that are analogous to the transition state usually bind strongly and potently inhibit enzyme activity by outcompeting the substrate [12].

Key Concept: Transition State Theory and Free Energy Diagrams

All chemical reactions pass through an unstable intermediate termed the transition state, which is a transitory structure between those of the substrates and products. The lifetime of the transition state is purported to be around 10^{-13} seconds, similar to the time for a single bond vibration. Although direct observation of the structure of the transition state is not possible, it is central to understanding catalysis, because enzymes function by lowering the activation energy barrier by tightly binding to the transition state. This allows a greater proportion of the substrate to reach the energy needed to proceed to the product.

Often a free energy diagram is constructed as a simple 2D attempt to mimic the 3D potential energy surface, which describes the set of potential energy paths the components of the reaction follow. The diagram shows the initial energy of the substrates, the transition state, and the products. The difference between the substrates' ground state and the transition state is the activation energy.

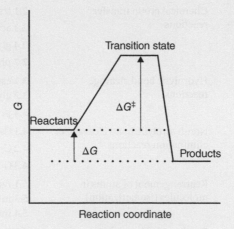

The energy required to progress from the reactant ground state to the transition state is called the activation energy (E_A) or energy barrier and is the difference in free energy between the reactant or ground state and the transition state. The height of the activation energy barrier is related

Figure 1.8 Energy profile diagram.
Reactants require energy (E_A) to reach their transition state and then form products. Products are formed through two routes: uncatalyzed (without enzyme, black) and catalyzed (with enzyme, red). The free energy of activation, $\Delta G^{\ddagger}$, is determined from the temperature dependence of the reaction rate, and the overall free energy change, ΔG, which is calculated from the reaction equilibrium constant.

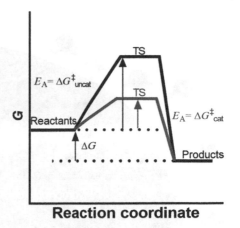

Reaction coordinate

to the rate of a chemical reaction. Enzymes work by reducing the size of this barrier and may increase the rates of reactions significantly (Figure 1.8).

Enzymes have unique structures that have evolved to enhance catalytic rates. The surface of the enzyme interacting with solvent water comprises mostly polar groups, whereas the interior of the protein is non-polar. Binding sites on the surface potentially allow for steep polarity gradients moving toward the interior of the protein, which provides the required environment for catalysis to occur. Additionally, hydrogen bonding, hydrophobic interactions, and charge–charge interactions involved in maintaining the enzyme's structure can be exploited to bind substrate molecules in a specific manner. In the absence of substrate, the binding site is filled with water molecules which must be displaced as the enzyme-substrate (ES) complex is formed. The energy of interaction must be great enough to expel this water and overcome the unfavorable entropy change associated with the formation of the complex. The reactivity is conferred by binding and introducing strain into the substrate so that it is biased towards the transition state. This ensures optimum binding of the transition state rather than the substrate itself, resulting in the exquisite alignment of the catalytic groups from the enzyme that will elicit the chemical step (Figure 1.9). This arrangement allows very large rate enhancements enabled by the stabilization of the transition state of the enzyme-catalyzed reaction relative to the uncatalyzed reaction. Enhancements in the rate of almost 10^{17}-fold have been observed in sweet potato β-amylase [13].

1.2.1.2 Specificity

Enzymes are incredibly specific, being able to discriminate between enantiomers (isomers of a molecule that are mirror images of each other), for example, the difference between L-amino acids (existing naturally) and D-amino acids and D-sugars (existing naturally) and L-sugars. This property also means only one stereoisomer of a drug may be capable of acting on the target; ibuprofen is an example. The active ingredient in ibuprofen is the S-isomer, which inhibits the cyclooxygenase enzyme (COX), but ibuprofen is a racemic mix of R and S isomers (Figure 1.10) [14, 15].

Enzymes are also able to catalyze the reactions they promote under surprisingly mild and economic conditions compared to other catalysts. Essentially, most enzymes seem to work best at temperatures close to body temperature, near neutral pH values, and at standard atmospheric pressures, qualities lacking in most industrial catalysts. For example, the Haber process is an

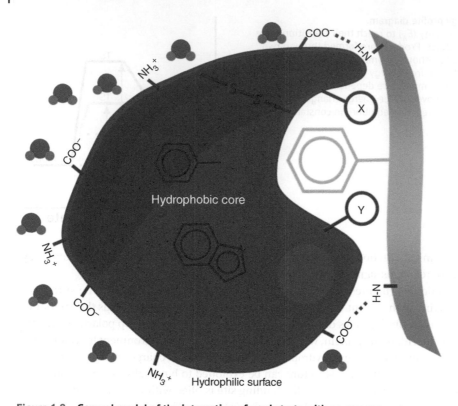

Figure 1.9   General model of the interaction of a substrate with an enzyme.
The hydrophobic core of the enzyme (purple) generally contains residues that have higher hydrophobicity. The hydrophilic surface contains residues, which interact with the solvent. Hydrogen bonding interactions are depicted between the enzyme and substrate (green), positioning the relevant substrate group close to the groups involved in catalysis (X and Y).

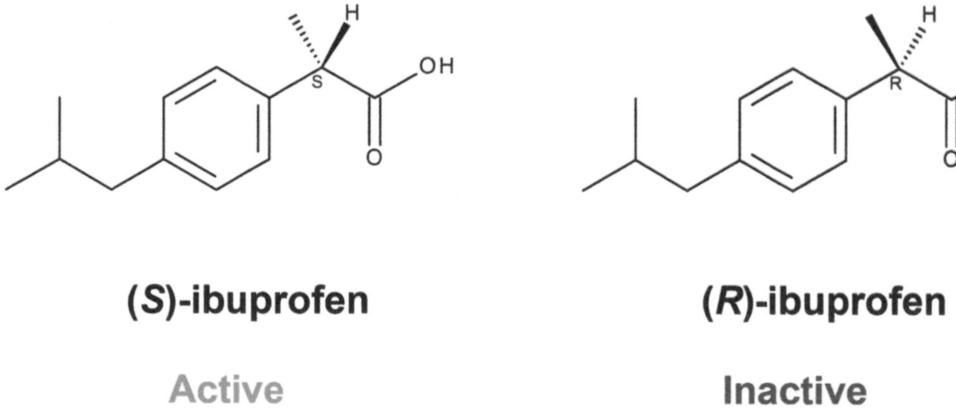

(S)-ibuprofen

Active

(R)-ibuprofen

Inactive

Figure 1.10   Ibuprofen stereoisomers.
Commercially bought ibuprofen is a mixture of stereoisomers; the *S*-isomer inhibits cyclooxygenase (COX), whereas the *R*-isomer is inactive against the enzyme.

industrial process using a catalyst, which produces ammonia. Iron is the catalyst used, which is relatively cheap and abundant, but high temperatures (400–450 °C) and high pressures (200 atm) are required.

1.2.1.3 Regulation

Enzymes may be regulated by other biological molecules, which can serve to enhance or decrease their catalytic ability. These biological molecules can be other proteins, like enzymes; preceding or subsequent enzymes in a signaling cascade can directly interact with a chosen enzyme to activate or deactivate it. An example is the rat sarcoma virus-rapidly accelerated fibrosarcoma (RAS-RAF) signaling pathway. Briefly, RAS is stimulated by an extracellular stimulus propagated through a receptor tyrosine kinase; RAF is recruited through RAS to the membrane and activated, it can then activate mitogen-activated extracellular signal-regulated kinase (MEK). MEK activates ERK and ERK can phosphorylate various regulatory proteins. Phosphorylation at specific sites on these enzymes can act to inhibit or activate their activity [16].

Excess substrate or product can also regulate enzyme activity. Generally, excess substrate can increase the activity of the enzyme, but excess substrate can also lead to substrate inhibition (Chapter 6). An example is phosphofructokinase (PFK), the enzyme responsible for phospho-rylating fructose-6-phosphate in glycolysis. An excess of ATP can bind to PFK and inhibit its activity, thus ensuring that glycolysis and subsequent metabolic processes do not occur when there is plenty of ATP [17, 18]. Accumulation of the product can also result in inhibition. Product inhibition (Chapter 6) can be observed in lipoxygenase (LOX) proteins such as ALOX15; increased concentrations of its product, 13-hydroxyoctadecadienoic acid, can inhibit enzyme activity even in the presence of its substrate, γ-linolenic acid [19].

Concentrations of biological molecules in cells are tightly regulated, and as a result, enzymatic activity is also tightly regulated. Dysregulation of specific enzymes to reduce or increase activity can result in particular disease states, such as cancer. For example, mutations of enzymes in a growth pathway can permanently activate them, and this can ultimately lead to tumorigenesis. Dysregulated enzymes can be targeted therapeutically to return activity to appropriate levels. As a result, enzyme regulation through small molecule invention is a large focus of pharmaceutical enzymology today [20, 21]. With this in mind, it is important to realize that the enzyme adopts multiple conformations during the catalytic cycle for the conversion of substrates to products. For example, considering an enzyme-catalyzed reaction involving two substrates and two products, the enzyme may be present as a free enzyme, or as a complex with either or both substrates or products (Figure 1.11). Recognizing which enzyme form is the most relevant and abundant for a particular disease process will facilitate drug design.

1.2.1.4 Stability

The environment of an enzyme can also alter enzyme activity; the presence of cofactors (e.g. ions), pH, and temperature can all regulate enzymatic activity. When these environmental factors are extreme, this can lead to enzyme denaturation. Denaturation is the unfolding of the tertiary structure of the enzyme into a disordered polypeptide in which key residues are no longer aligned and therefore unable to form key functional and stabilizing interactions, resulting in a concomitant loss of catalytic activity. Unfolding of the protein is often accompanied by kinetically irreversible steps such as covalent changes or aggregation, meaning that enzyme denaturation is often an irreversible

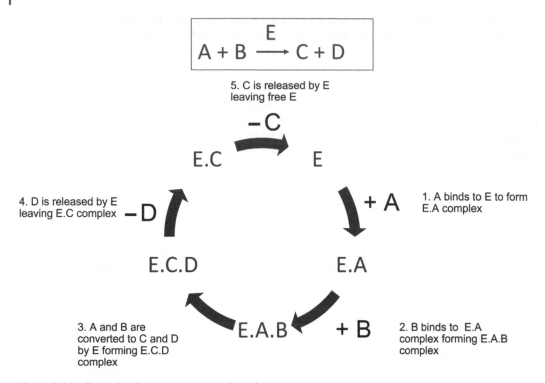

Figure 1.11   Example of an enzyme catalytic cycle.
A and B are substrates, one substrate binds before the other and then they are converted into two products C and D. Product D is then released before product C. The cycle begins again after the final product is released. Each of the different enzyme forms can be targeted for drug discovery if related to a particular disease state.

process. Refolding of the protein can be attempted; however, this often results in an enzyme with lesser or no catalytic activity.

Denaturation of an enzyme can occur by physical or chemical means. Each enzyme has a specific set of physical conditions for optimal catalytic activity and even small changes in these can lead to reduced enzyme activity, and eventually denaturation. Amongst, the most critical conditions for enzyme function are temperature and pH. Whilst most enzymes function optimally at temperatures close to 37 °C and near-neutral pH, enzymes which are located in more extreme environments have adapted to function optimally under these conditions. For example, the protease pepsin which is located in the stomach functions optimally at ~pH 2 [22], and the Taq polymerase which originates from the thermophilic bacterium *Thermus aquaticus* functions optimally at ~75 °C [23]. Chemical denaturation occurs when enzymes are exposed to chemical agents, which cause the protein to unfold. Examples of these are solvents, chaotropic agents such as urea, and chemically reactive agents such as hydrogen peroxide.

Biophysical techniques such as differential scanning fluorimetry (DSF) and circular dichroism (CD) assess protein unfolding/folding and can be employed to check that a protein is correctly folded. DSF measures protein unfolding by monitoring changes in fluorescence as a function of temperature (this can be intrinsic protein fluorescence or a hydrophobic dye that binds to the exposed hydrophobic residues of a protein as it unfolds), whilst CD measures protein folding and the formation of secondary structures by monitoring the absorption of circularly polarized light. Aggregation of proteins can also be measured biophysically using techniques such as dynamic light scattering (DLS) and multi-angle light scattering (MALS), which detect the size distribution profile of macromolecular structures and can provide information on the oligomeric state of a protein.

References

1 Berg JM, Stryer L, Tymoczko JL, Gatto GJ. *Biochemistry*: Macmillan Learning; 2015.

2 Nelson DL, Lehninger AL, Cox MM. *Lehninger Principles of Biochemistry*: W. H. Freeman; 2008.

3 Regad L, Martin J, Nuel G, Camproux AC. Mining protein loops using a structural alphabet and statistical exceptionality. *BMC Bioinformatics*. 2010;11:75.

4 Fiser A, Sali A. ModLoop: automated modeling of loops in protein structures. *Bioinformatics (Oxford, England)*. 2003;19(18):2500–1.

5 Wojcik J, Mornon JP, Chomilier J. New efficient statistical sequence-dependent structure prediction of short to medium-sized protein loops based on an exhaustive loop classification. *Journal of Molecular Biology*. 1999;289(5):1469–90.

6 Sun PD, Foster CE, Boyington JC. 17 – Overview of Protein Structural and Functional Folds. *Current Protocols in Protein Science*: Wiley; 2004; Unit 17.1 35: 17.1.1–17.1.189. https://doi.org/10.1002/0471140864.ps1701s35

7 Perrakis A, Sixma TK. AI revolutions in biology: the joys and perils of AlphaFold. *EMBO Reports*. 2021;22(11):e54046.

8 Jumper J, Evans R, Pritzel A, Green T, Figurnov M, Ronneberger O, et al. Highly accurate protein structure prediction with AlphaFold. *Nature*. 2021;596(7873):583–9.

9 Callaway E. AlphaFold's new rival? Meta AI predicts shape of 600 million proteins. *Nature*. 2022;611(7935):211–2.

10 Raimondi MV, Randazzo O, La Franca M, Barone G, Vignoni E, Rossi D, et al. DHFR inhibitors: reading the past for discovering novel anticancer agents. *Molecules*. 2019;24(6):1140.

11 Yocum RR, Rasmussen JR, Strominger JL. The mechanism of action of penicillin. Penicillin acylates the active site of *Bacillus stearothermophilus* D-alanine carboxypeptidase. *Journal of Biological Chemistry*. 1980;255(9):3977–86.

12 McDonald A, Tipton K. Enzyme nomenclature and classification: the state of the art. *The FEBS Journal*. 2021, 290 2214–31.

13 Wolfenden R, Lu X, Young G. Spontaneous hydrolysis of glycosides. *Journal of the American Chemical Society*. 1998;120(27):6814–5.

14 Evans AM. Comparative pharmacology of *S*(+)-ibuprofen and (*RS*)-ibuprofen. *Clinical Rheumatology*. 2001;20(1):9–14.

15 Dalziel K. Dynamic aspects of enzyme specificity. *Philosophical Transactions of the Royal Society of London Series B, Biological Sciences*. 1975;272(915):109–22.

16 Chong H, Lee J, Guan KL. Positive and negative regulation of Raf kinase activity and function by phosphorylation. *EMBO Journal*. 2001;20(14):3716–27.

17 Cabrera R, Ambrosio ALB, Garratt RC, Guixé V, Babul J. Crystallographic structure of phosphofructokinase-2 from *Escherichia coli* in complex with two ATP molecules. Implications for substrate inhibition. *Journal of Molecular Biology*. 2008;383(3):588–602.

18 Fenton AW, Reinhart GD. Mechanism of Substrate inhibition in *Escherichia coli* phosphofructokinase. *Biochemistry*. 2003;42(43):12676–81.

19 Joshi N, Hoobler EK, Perry S, Diaz G, Fox B, Holman TR. Kinetic and structural investigations into the allosteric and pH effect on the substrate specificity of human epithelial 15-lipoxygenase-2. *Biochemistry*. 2013;52(45):8026–35.

20 Holdgate GA, Meek TD, Grimley RL. Mechanistic enzymology in drug discovery: a fresh perspective. *Nature Reviews Drug Discovery.* 2018;17(2):115–32.

21 Imming P, Sinning C, Meyer A. Drugs, their targets and the nature and number of drug targets. *Nature Reviews Drug Discovery.* 2006;5(10):821–34.

22 Piper DW, Fenton BH. pH stability and activity curves of pepsin with special reference to their clinical importance. *Gut.* 1965;6(5):506–8.

23 Chien A, Edgar DB, Trela JM. Deoxyribonucleic acid polymerase from the extreme thermophile *Thermus aquaticus. Journal of Bacteriology* 1976;127(3):1550–7.

2

Binding Equilibria and Kinetics

CHAPTER MENU

2.1 Introduction to Chemical Kinetics

Chemical kinetics allow the understanding of biological processes, especially enzyme-catalyzed reactions, and are centered upon the measurement of the rate of reaction, or simply how quickly reactions happen. By modifying the reaction conditions, the effect of these changes can be used to interrogate the reaction at the molecular level. Chemical reactions can vary widely in the speed at which they occur, with the dynamic range from milli-seconds to several years to reach equilibrium. The reaction rate is the change in the concentration of reactants or products over time.

There are many ways to measure the change in concentration of these species with time, and different methods are applied according to the timescales involved. However they are measured, there are several different types of reactions that can be classified according to how the rate changes with the concentration of reactants.

2.1.1 Zero-order Reactions

In some cases, the rate of reaction appears to be independent of reactant concentration. Thus, the rate does not change when the concentration of reactant is increased or decreased.

Consider a reaction in which the rate of change for reactant A is constant. For this zero-order reaction, the rate of change (v) of concentration of A ($[A]$) is given by Equation 2.1 and is illustrated in Figure 2.1

$$v = -\frac{d[A]}{dt} = k[A]^0 = k$$

Equation 2.1 **Expression for the rate equation for a zero-order reaction. Note the rate is independent of concentration.**

Laboratory Guide to Enzymology, First Edition. Geoffrey A. Holdgate, Antonia Turberville, and Alice Lanne.
© 2024 John Wiley & Sons, Inc. Published 2024 by John Wiley & Sons, Inc.

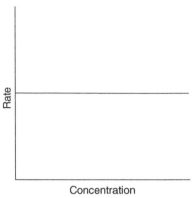

Figure 2.1   Plot of rate versus concentration for a zero-order reaction.
This plot shows that the rate of a zero-order reaction is independent of concentration.

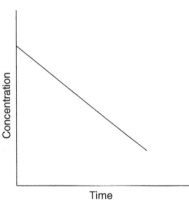

Figure 2.2   Plot of concentration versus time for a zero-order reaction.
This figure highlights that the reactant concentration changes linearly with time for a zero-order reaction.

The integrated form of the rate equation allows the concentration of the reactant to be calculated at any time (t) during the reaction. The integrated form of a zero-order reaction is given by Equation 2.2, where $[A]_0$ is the concentration of A at $t = 0$. The change in concentration of reactant over time is illustrated in Figure 2.2.

$$[A] = [A]_0 - kt$$

Equation 2.2   Expression for the integrated rate equation for a zero-order reaction.

In this case, the rate constant (k) has the units of Ms^{-1}. This situation may occur early in the enzyme-catalyzed reaction time course if the enzyme concentration is very low compared to the substrate concentration and so the reaction may appear to be zero order. Of course, a zero-order process cannot continue after the reactant has been completely depleted.

2.1.2   First-order Reactions

Where the rate depends linearly on the concentration of one reactant (A), the reaction is termed first order. The rate equation is given by Equation 2.3, the behavior is illustrated in Figure 2.3 and the rate constant has the units s^{-1}.

$$v = -\frac{d[A]}{dt} = k[A]$$

Equation 2.3   Expression for the rate equation for a first-order reaction. Note the rate is dependent upon the reactant concentration.

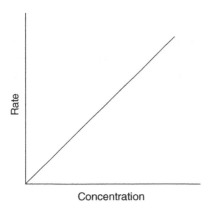

Figure 2.3 Rate versus concentration plot for a first-order reaction.
This plot demonstrates that the rate of reaction is directly proportional to concentration of reactant for a first-order reaction.

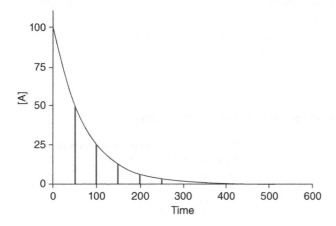

Figure 2.4 Exponential decay of reactant A.
As time proceeds the concentration of reactant A ([A]) will decrease, in a first-order reaction the time taken for it to half ($t_{1/2}$) is consistent. The first five half-lives are shown in red and [A] halves every 50-time units.

The integrated rate equation is given in Equation 2.4.

$$[A] = [A]_0 e^{-kt}$$

Equation 2.4 Expression for the integrated rate equation for a first-order reaction.

It can be seen that the change in the concentration of A follows an exponential relationship and from this, it can be shown that the time taken for the concentration to drop to half of its value is a constant, regardless of the initial concentration of A (Figure 2.4). This time is termed the half-life, $t_{1/2}$, and is related to the rate constant, k by Equation 2.5.

$$t_{1/2} = \frac{\ln 2}{k}$$

Equation 2.5 Expression illustrating the relationship between half-life and the first-order rate constant, *k*.

2.1.3 Second-order Reactions

In second-order reactions, the sum of the exponents in the rate equation is equal to 2. There are two common examples of second-order reactions. The first is where two molecules of the same reactant combine to yield a single product (Reaction scheme 2.1) and the rate is given by Equation 2.6 and has the units $M^{-1}s^{-1}$.

$$2A \longrightarrow P$$

Reaction scheme 2.1 **Reaction scheme for the reaction between two identical reactants combining to form a single product.**

$$v = -\frac{d[A]}{dt} = k[A]^2$$

Equation 2.6 **Expression for the rate equation for a second-order reaction. Note the rate is dependent upon both reactant concentrations, which here happen to be identical.**

As is shown in Equation 2.6, the rate is directly proportional to the concentration of reactant squared, yielding a plot of the form shown in Figure 2.5.

The integrated rate equation for this second-order reaction is given by Equation 2.7. The change in concentration with time is illustrated in Figure 2.6.

$$\frac{1}{[A]} = \frac{1}{[A]_0} + kt$$

Equation 2.7 **Expression for the integrated rate equation for a second-order reaction.**

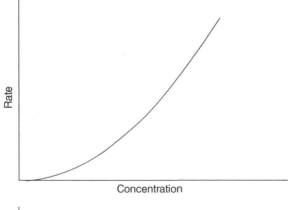

Concentration

Figure 2.5 **Rate versus concentration plot for a second-order reaction.**
This figure shows the increase in rate with concentration of reactant for a second-order reaction. The rate is proportional to the concentration of reactant squared.

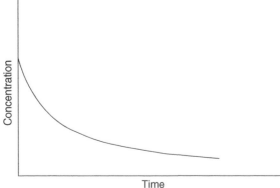

Time

Figure 2.6 **Concentration versus time plot for a second-order reaction.**
This plot shows that the concentration of reactant decreases with time for a second-order rate constant. There is not a constant value of half-life as this depends upon the initial concentration of reactant.

A half-life can also be defined for this second-order reaction by Equation 2.8.

$$t_{1/2} = \frac{1}{k[A]_0}$$

Equation 2.8   Expression illustrating the relationship between half-life and the second-order rate constant, k.

In this case, the half-life is dependent upon the initial concentration of the reactant, and so may have less utility than that for a first-order reaction. The second common example of a second-order reaction is where two different molecules combine to yield a single product (Reaction scheme 2.2). The rate is given by Equation 2.9.

$$A + B \longrightarrow P$$

Reaction scheme 2.2   Reaction scheme for the reaction between two different reactants combining to form a single product.

$$v = -\frac{d[A]}{dt} = k[A][B]$$

Equation 2.9   Expression for the rate equation for a second-order reaction. Note the rate is dependent upon both reactant concentrations, which in this case are different.

The integrated rate equation is given by Equation 2.10.

$$\ln \frac{[A]_0[B]}{[B]_0[A]} = k([B]_0 - [A])t$$

Equation 2.10   Expression for the integrated rate equation for a second-order reaction.

2.1.4   Pseudo-First-Order Reactions

Under certain conditions, it may be possible to simplify the second-order rate equation, such that it approximates to a first-order rate equation. Consider the situation when, in the reaction of A and B to yield P (Reaction scheme 2.2), the initial concentration of B ($[B]_0$) is much greater than the initial concentration of A ($[A]_0$), then [B] can be considered to equal $[B]_0$, so that Equation 2.10 above becomes Equation 2.11.

$$\ln \frac{[A]_0}{[A]} = k[B]t$$

which can be rearranged to give:

$$[A] = [A]_0 e^{-[B]kt}$$

Equation 2.11   Expression for the integrated rate equation for a pseudo-first-order reaction.

This has the same form as the integrated first-order rate equation above (Equation 2.4). Assuming that [B] does not change during the reaction, an apparent rate constant (k') can be used to replace $[B]k$ and so the rate becomes the equation below (Equation 2.12) and so, under these conditions, the second-order reaction can be treated as if it were first order.

$$v = k'[A]$$

Equation 2.12   Expression for the rate equation for a pseudo-first-order reaction.

2.1.5 Temperature Dependence of Rate Constants

The rates of chemical reactions vary with temperature, where generally an increase in temperature leads to an increase in the rate of reaction. This is because most chemical reactions occur through molecular collisions, whose frequency increase with increasing temperature. Alongside increased collisions, the molecules will have higher kinetic energy, increasing the probability that a collision will result in a productive reaction.

The Eyring equation (Equation 2.13) can be used to describe the change in the rate constant with temperature.

$$k = \frac{k_{\mathrm{B}}T}{h}e^{\left(\frac{\Delta S}{R} - \frac{\Delta H}{RT}\right)}$$

Equation 2.13 **Eyring equation, describing the variation of the value of a rate constant with temperature.**

where k_{B} is the Boltzmann constant, h is the Plancks constant, T is the absolute temperature, R is the gas constant, ΔH is the enthalpy of activation, and ΔS is the entropy of activation. This simplified version of the Eyring equation assumes that the transmission coefficient, k, has a value of 1 (which assumes that species reaching the transition state always proceed to products, rather than returning to reactants).

2.2 Introduction to Binding Kinetics

Enzymes can bind multiple different ligands, including substrates, cofactors, and small molecules. In the presence of a ligand (L), the macromolecule (M) will be in at least two different states (free and ligand-bound), and the balance between these states is known as an equilibrium.

The equilibrium consists of two reactions: the forward reaction ($k_{\mathrm{on}}[\mathrm{M}][\mathrm{L}]$) where the free macromolecule binds to the ligand to form the macromolecule-ligand complex (ML) and the reverse reaction ($k_{\mathrm{off}}[\mathrm{ML}]$) where the enzyme-ligand complex dissociates back into its unbound forms (Figure 2.7).

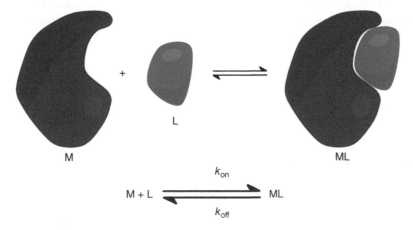

Figure 2.7 **Binding of a ligand to a macromolecule.**
Unbound macromolecule (M) associates with free ligand (L) to form the complex (ML), the association rate constant is k_{on}. The complex can dissociate back into the unbound forms, the dissociation rate constant is k_{off}.

The forward reaction rate is a combination of the concentrations of the free macromolecule and ligand and the rate constant, k_{on}. k_{on} is the rate constant for this interaction and is independent of concentration. The reverse reaction rate is dependent on the concentrations of the bound complex ([ML]) and k_{off} is the rate constant. Each reaction proceeds at different rates and early principles describing the relationship between these rates were defined in 1879 by Guldberg and Waage [1–3]. Their work led to two important concepts:

- A reversible reaction reaches a state of dynamic equilibrium.
- The law of mass action.

Key Concept: Dynamic Equilibrium

A reversible reaction is in dynamic equilibrium when the forward and reverse reactions proceed at equal rates. This assumes that the reactant and product concentrations do not change and that the reaction occurs in a closed system.

There are many ways to investigate the interaction between enzyme, substrate, and small molecules and in this chapter, we will focus on isolated enzyme assays. Other methods to investigate binding interactions not covered here include structure determination using X-ray crystallography [4] or nuclear magnetic resonance (NMR) [5, 6], site-directed mutagenesis [7], ultracentrifugation [8, 9], spectroscopic techniques and other methods (e.g. mass spectrometry) [10].

Key Concept: The Law of Mass Action

The law of mass action states that the rate of a reaction is proportional to the active masses of the reactants, expressed in mol/L. This assumes that the active mass is equal to the concentration × the activity coefficient, often assumed to be close to 1.

2.3 Ligand Binding to Single Binding Site

Ligands can bind to different sites on a macromolecule. The simplest mechanism is where the ligand binds to a single site, for example, when an inhibitor binds to the active site of an enzyme (we are taking this case as the binding of substrate would lead to catalysis and so complicates the analysis of binding). The affinity of the interaction, or in other words the stability of the ML complex is a ratio of the association and dissociation rate constants and is defined by K_d, the equilibrium dissociation constant, or K_a, the association constant. These constants can be calculated from the rate constants k_{on} and k_{off} or concentrations of unbound macromolecule and ligand and ML complex (Equations 2.14 and 2.15).

$$K_d = \frac{[M][L]}{[ML]} = \frac{k_{off}}{k_{on}}$$

Equation 2.14 **Expression for the equilibrium dissociation constant, K_d from rate constants.**

$$K_a = \frac{[ML]}{[M][L]} = \frac{k_{on}}{k_{off}} = \frac{1}{K_d}$$

Equation 2.15 **Expression for the equilibrium association constant, K_a from rate constants.**

The equilibrium dissociation constant (K_d) is more frequently used compared to the K_a when describing affinity as the units for this term is concentration (e.g. M, mM, μM) rather than inverse concentration units (e.g. M^{-1}, mM^{-1}, $μM^{-1}$) used for the equilibrium association constant. The K_d represents the free ligand concentration when 50% of the macromolecule sites are occupied with ligand, that is [M] = [ML]. Smaller K_d values indicate a stronger interaction between the macromolecule and ligand and increased stability of the ML complex.

Key Concept: Rate Constant and Equilibrium Constant Nomenclature

Equilibrium constants are defined by a K (e.g. K_d) and are ratios of rate constants, which are defined by a k (e.g. k_{on}).

A plot of [ML] or a signal proportional to [ML], plotted versus [L] can be used to estimate K_d (Figure 2.8) using non-linear regression, by fitting to the Langmuir binding isotherm (Equation 2.16).

$$[ML] = \frac{[M]_t[L]}{K_d + [L]} = \frac{[M]_t}{1 + \frac{K_d}{[L]}} = [B] = \frac{B_{max}[L]}{K_d + [L]}$$

Equation 2.16   Expression for the Langmuir binding isotherm.

In the Langmuir binding isotherm, [B] refers to the concentration of ligand-bound and B_{max} is the maximum possible concentration of ligand-bound, which is equivalent to the total concentration of binding sites (Figure 2.8).

Sometimes linear transformations of the Langmuir binding isotherm are used to display data. This can be useful, but linear regression should not be used to estimate binding parameters since the normal error distribution may be skewed or uneven weighting introduced by the transformation, see Chapter 9 for more detail.

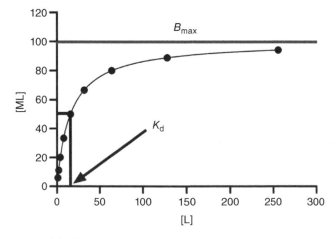

Figure 2.8   Binding of a ligand to a single site on a macromolecule.
A plot of the concentration of macromolecule bound to ligand, [ML] against the concentration of free ligand, [L] allows determination of the maximum bound macromolecule concentration, $B_{max,}$ and calculation of the concentration at which 50% of the macromolecule is bound to ligand, also known as the equilibrium dissociation constant and K_d.

Key Concept: The Langmuir Binding Isotherm

The Langmuir binding isotherm was derived by Irving Langmuir in 1916 to describe the adsorption of gas molecules to a solid surface at a constant temperature, these equations have been applied to ligand binding. In the context of ligand binding the Langmuir binding isotherm can measure the fractional occupancy of a macromolecule and calculate K_d.

The assumptions for the Langmuir binding isotherm are:

- [M] is $<<< K_d$ and therefore the free ligand concentration [L] is equal to total ligand concentration, $[L]_t$
- There is only one ligand binding site per macromolecule and there is no interaction between molecules at another site.

The procedure of binding is a dynamic process, with M and L associating to form the ML complex, but with the reverse process also occurring with some of the ML complexes dissociating back to free M and L. This dynamic process is still occurring at equilibrium; however, the rate of association is now equal to the rate of dissociation.

2.3.1 Specific Versus Non-specific Binding

Ligand binding to a macromolecule can be specific or non-specific (Figure 2.9). Specific binding means the ligand binds to the macromolecule in a biologically relevant way, for example, the ligand binds to the active site of an enzyme or a known regulatory site. Non-specific binding to a macromolecule is binding that is not biologically relevant, the ligand can bind anywhere on the macromolecule and can often interfere with analysis of the binding affinity. In cases where there is non-specific binding, the binding response should be corrected for non-specific binding before the data are analyzed using (Equation 2.17).

$$[ML]_{spec} = [ML]_t - [ML]_{nonspec}$$

Equation 2.17 **Expression for the specific component of binding relative to total binding.**

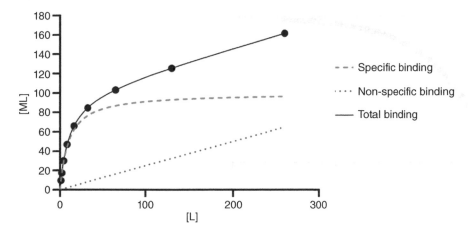

Figure 2.9 **Specific versus non-specific binding.**
A plot of the concentration of macromolecule bound to ligand, [ML] against the concentration of free ligand, [L]. To gain an accurate binding affinity, total binding and non-specific binding are measured and the non-specific binding is subtracted from the total binding to give specific binding.

2.4   Kinetic Approach to Equilibrium

The approach to equilibrium with time can often be described kinetically using a pseudo-first-order rate constant, k_{obs} (observed rate constant), since the association often occurs with little change in [L] (free ligand), and $[L]_t$ (total ligand) is much greater than $[M]_t$ (total macromolecule) and so $[L] \approx [L]_t$. The reaction proceeds with pseudo-first-order kinetics and is shown by Equation 2.18.

$$[ML]_t = [ML]_{max}\left\{1 - e^{(-k_{obs}t)}\right\} = [B]_t = B_{max}\left\{1 - e^{(-k_{obs}t)}\right\}$$

Equation 2.18   Expression describing the approach to equilibrium for the pseudo-first-order binding of a ligand to a macromolecule.

k_{obs} can be estimated by plotting $[ML]_t$ (where $[ML]_t$ or $[B]_t$ is the bound ligand concentration at time t), or a signal proportional to it versus time (Figure 2.10).

By determining k_{obs} at various $[L]_t$($\approx [L]$) the values of k_{on} and k_{off} can also be estimated, using linear regression. The value of k_{obs} is given by Equation 2.19.

$$k_{obs} = k_{off} + k_{on}[L]$$

Equation 2.19   Expression showing relationship between k_{obs} and the association and dissociation rate constants.

The values of these rate constants are variable, with the upper limit for the association rate constant being governed by the rate of diffusion of the interacting molecules, with an upper limiting value of $1 \times 10^7 - 1 \times 10^8$ M^{-1} s^{-1}. Lower values of the association rate constant lead to slow binding, although the rate of binding can be increased by increasing the concentration of the interacting partners (Figure 2.11). The dissociation rate constant also varies, with the ratio of the two governing K_d (see Section 2.2).

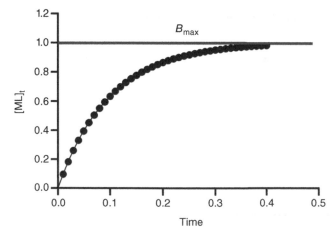

Figure 2.10   Approach to equilibrium.
A plot of the concentration of macromolecule bound to ligand, $[ML]_t$ against time, allows determination of the maximum bound macromolecule concentration, $B_{max,}$ and determination of the pseudo-first-order rate constant, k_{obs}

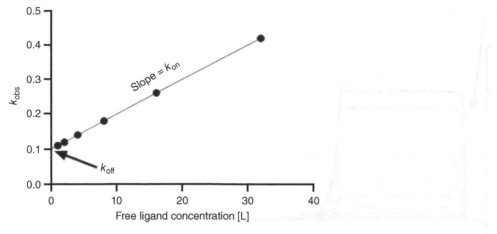

Figure 2.11 **Evaluation of binding rate constants (k_{on} and k_{off}).**
A plot of k_{obs} versus [L] allows determination of the values of k_{on} and k_{off} as the slope and intercept of the plot, respectively.

2.5 Methods for Measuring Protein-ligand Binding

There are different methods used to measure protein-ligand binding, the most common methods used are surface plasmon resonance (SPR) and isothermal titration calorimetry (ITC).

A requirement of SPR experiments is that one binding partner, termed the ligand, is attached to the biosensor surface. There are several approaches by which this immobilization may be achieved, including either covalent or non-covalent methods. Usually, the protein is attached to the hydrogel matrix, and this must be accomplished in a manner which retains functional binding activity and capacity.

Once a suitable ligand surface has been prepared, the analyte is injected over the ligand to monitor binding. Binding occurs when the analyte associates with the immobilized ligand, following the diffusion of analyte from bulk flow to the sensor surface.

As dissociation events also occur during the injection of analyte, if the injection of analyte continues for long enough, then the net binding rate observed will fall to zero (shown when the binding response becomes parallel to the baseline), as the rate of dissociation becomes equal to the rate of association. This situation is called steady-state and reflects the fact that the concentration of ligand-analyte complex is not changing.

After binding has taken place, the ligand and analyte remain bound in the complex until dissociation occurs. Dissociation is the process by which the analyte is released from the complex and is removed from the sensor surface by buffer flow. When the flow of analyte is stopped and replaced with buffer, the concentration of the free analyte falls to zero, and dissociation is detected as a decrease in signal.

The SPR sensorgram allows measurement of the kinetics and affinity of the interaction. Typically, the concentration of ligand is varied above and below the estimated K_d (Figure 2.12). By fitting curves to this data, the R_{max} (maximum bound), k_{on}, k_{off}, and K_d can be calculated. SPR experiments can also be used for ligand competition studies (described in Section 2.8). A standardized protocol (Protocol 3.6) can be found in Chapter 3.

ITC is the only technique which directly measures the enthalpy change for a bimolecular binding interaction (Figure 2.13). The sample cell contains the macromolecule, into which the ligand is

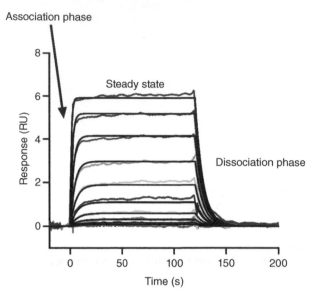

Figure 2.12   Surface plasmon resonance sensorgram.
SPR sensorgram showing association and dissociation phases, for binding of different concentrations of ligand to an immobilized macromolecule. The signal increases during the association phase, as the concentration of the complex increases to the equilibrium concentration. During the dissociation phase, the instrument passes buffer over the complex, which dissociates, leading to a decrease in signal as the mass at the sensor surface decreases.

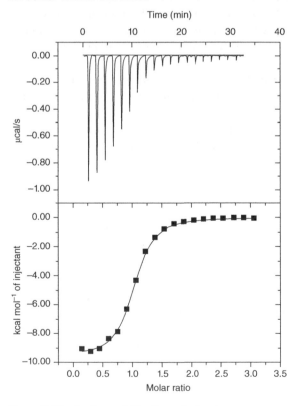

Figure 2.13   Isothermal titration calorimetry.
Isothermal titration calorimetric thermogram showing binding of a ligand to a macromolecule.

titrated. A reference cell contains buffer. The heat effects arising from binding in the sample cell cause a temperature difference (ΔT) between the two cells, which is detected by the calorimeter and triggers a change in the feedback power applied to the sample cell (power compensation). Exothermic reactions produce a decrease in the applied power, whilst endothermic reactions produce increased feedback. The change in feedback power applied to the sample cell with respect to the experimental time is measured and represents the experimental raw data or thermogram. The heat associated with the binding interaction is obtained by integration of the peaks of the power versus time plot, with respect to time.

The ITC experiment often can be designed to enable estimations for enthalpy, affinity, and stoichiometry within a single experiment. Most of the added ligand becomes bound during the initial injections, due to the excess of macromolecule, and gives a model-independent estimate of $\Delta H°$. The heat change decreases throughout the titration as the available macromolecule sites are filled, and some of the added ligand remains free in solution. This part of the titration curve allows estimation of affinity. The stoichiometry is measured, assuming the concentrations of functional macromolecule and ligand are accurately known as the molar ratio of ligand bound to macromolecule at the equivalence point (where bound [macromolecule] = free [macromolecule]).

These three parameters (including K_d) describing the interaction are calculated by fitting the appropriate equation to the integrated heat data. A protocol (Protocol 3.5) can be found in Chapter 3.

2.6 Ligand Depletion (Tight Binding)

If the affinity of the ligand for the macromolecule is high, then the varied ligand concentrations will be similar to the concentration of macromolecule, and [L] will no longer be approximately $[L]_t$. Instead, depletion of [L] occurs by binding to the macromolecule (Figure 2.14). This situation is common for compounds with high affinity and/or when the concentration of enzyme binding

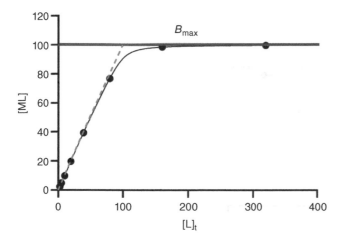

Figure 2.14 Tight binding of ligand to macromolecule.
High-affinity ligands can exhibit tight-binding behavior, the assumption that the unbound ligand concentration, [L] is equivalent to the total ligand concentration, $[L]_t$, no longer applies. When tight binding occurs, [L] becomes depleted as it binds to the macromolecule. This can be identified by very steep [ML] versus $[L]_t$ plots and a K_d that is dependent on the macromolecule concentration, [M]. When a tangent is drawn from the origin, the concentration at which it insects the B_{max} asymptote, gives an indication of [M].

sites present in the assay is high relative to this affinity. In this situation, another equation must be used to describe the binding, which in this case, is described as tight binding (Equation 2.20).

$$[ML] = \frac{[M]_t + [L]_t + K_d - \sqrt{([M]_t + [L]_t + K_d)^2 - 4([M]_t[L]_t)}}{2}$$

Equation 2.20 **General expression for binding of a ligand to a macromolecule, valid when ligand depletion occurs (tight-binding equation).**

2.7 Ligand Binding to Multiple Binding Sites

Proteins sometimes may have more than one binding site for the same ligand. These sites can be identical or non-identical and may function independently or binding at one site may influence the binding at subsequent sites.

2.7.1 Identical Independent Binding Sites

Identical binding sites require that the binding constants for ligand binding to each site are the same. When binding of a ligand to an independent identical site on the protein occurs, it does not affect binding at the other sites. As a result, the binding process may be described by Equation 2.21, where n is the number of binding sites.

$$[ML] = \frac{n[M]_t}{1 + \frac{K_d}{[L]}} = \frac{n[M]_t[L]}{[L] + K_d}$$

Equation 2.21 **Expression for ligand binding to a macromolecule having n identical, independent binding sites.**

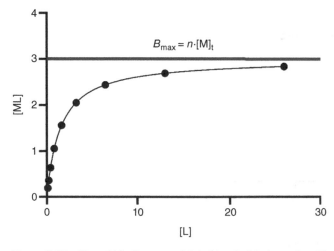

Figure 2.15 **Ligand binding to multiple identical independent binding sites**.
Where a protein has multiple identical, independent binding sites, ligands will bind to these with the same affinity. The maximum binding will be a product of the macromolecule concentration and the number of binding sites. In the example above the number of binding sites is three and the protein concentration is equal to 1.

This holds since it makes no difference whether the binding occurs at a macromolecule with only one binding site, or whether n sites are available on the same macromolecule. Although the equation has a similar form to the one-site binding equation (Equation 2.16), with the number of binding sites shown in the numerator, it should be noted that [B] can no longer be equated to [ML] (Figure 2.15). This is because [B] comprises all the partially saturated forms of the protein. For example, consider a protein with n identical independent binding sites, then [B] is given by Equation 2.22, and the individual sites on the protein are filled sequentially (Reaction scheme 2.3).

$$[B] = [ML] + 2[ML_2] + 3[ML_3] + \cdots + n[ML_n]$$

Equation 2.22 **Expression relating total ligand-bound macromolecule and each partially saturated bound form.**

$$M + L \rightleftharpoons ML$$

$$ML + L \rightleftharpoons ML_2$$

$$ML_2 + L \rightleftharpoons ML_3$$

$$ML_{n-1} + L \rightleftharpoons ML_n$$

Reaction scheme 2.3 **Sequential occupancy of multiple identical independent binding sites.**

All the individual macroscopic dissociation constants are equal and are given by Equation 2.23.

$$K_1' = \frac{[M][L]}{[ML]} = K_2' = \frac{[ML][L]}{[ML_2]} = K_3' = \frac{[ML_2][L]}{[ML_3]} = K_n' = \frac{[ML_{n-1}][L]}{[ML_n]}$$

Equation 2.23 **Expression relating the individual macroscopic dissociation constants for a macromolecule with n identical, independent binding sites.**

2.7.2 Non-identical Independent Binding Sites

Where the protein has multiple non-identical binding sites, binding of the ligand will occupy the site with the highest affinity first. Occupation of the sites with lower affinity will require higher ligand concentrations. If the sites are independent, then each different binding site is saturated according to Equation 2.21 above, with the total binding described by the sum of all the individual binding site saturation functions (Equation 2.24).

$$[ML] = \frac{n_1[M]_t}{1 + \frac{K_{d1}}{[L]}} + \frac{n_2[M]_t}{1 + \frac{K_{d2}}{[L]}} + \frac{n_3[M]_t}{1 + \frac{K_{d3}}{[L]}} + \cdots + \frac{n_m[M]_t}{1 + \frac{K_{dm}}{[L]}}$$

or

$$[ML] = \frac{n_1[M]_t[L]}{[L] + K_{d1}} + \frac{n_2[M]_t[L]}{[L] + K_{d2}} + \frac{n_3[M]_t[L]}{[L] + K_{d3}} + \cdots + \frac{n_m[M]_t[L]}{[L] + K_{dm}}$$

Equation 2.24 **Expression for ligand binding to a macromolecule having m different binding sites with n sites of the same affinity.**

Here, the different K_d values represent the dissociation constants for the different binding sites. m represents the number of different binding sites (that is with different affinity), and n represents the number of individual sites having the same affinity.

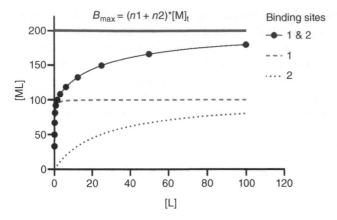

Figure 2.16 **Binding of multiple non-identical independent binding sites.**
Where a protein has multiple, non-identical binding sites, ligands will bind to these sites with different affinities. The tighter binding sites will be filled at lower ligand concentrations, with the weaker sites being saturated at higher concentrations of ligand. The binding isotherm will be a composite of the binding to each individual site – the addition of the dashed and dotted lines to yield the solid line. At saturation of all sites, the B_{max}, will reflect the total protein concentration and the number of binding sites.

It is clear that the resulting binding curve is a combination of the individual hyperbolic curves, as demonstrated below for the situation with just two non-identical independent binding sites (Figure 2.16).

The resulting curve is not purely hyperbolic, having a more prominent increase at lower ligand concentrations, as the highest affinity sites are preferentially filled. However, sometimes it can be difficult to distinguish this curve shape from a standard hyperbolic plot, especially if the affinities for different sites are relatively similar. This highlights the importance of high-quality data, utilizing different models, and bringing an understanding of the system into the data analysis.

2.7.3 Cooperativity

A cooperative binding model, to describe the behavior of oxygen binding to hemoglobin, was first proposed by Hill over 100 years ago but is still used widely today to describe cooperative behavior for protein-ligand binding (Equation 2.25).

$$[B] = \frac{[M]_t[L]^n}{[L]^n + K_d{}^n}$$

Equation 2.25 **Expression for the Hill equation describing the relationship for cooperative binding of a ligand to a macromolecule.**

In Equation 2.25, n is the Hill coefficient. When $n > 1$, the system shows positive cooperativity, which means that the sites interact such that filling the first site causes an increase in affinity at the next. Conversely, when $n < 1$, the system shows negative cooperativity, and binding at one site leads to a decrease in affinity at another. When positive cooperativity occurs, the binding curve is sigmoidal (Figure 2.17). When $n = 1$, it is easy to see that the equation reduces to Equation 2.16, where there is no cooperativity, and the sites are independent. Negative cooperativity produces a

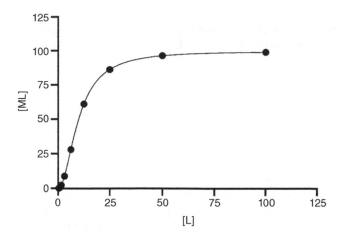

Figure 2.17 Cooperative binding.
Sigmoidal binding curve obtained for a compound displaying positive cooperativity. Here $n = 2$.

biphasic curve shape, which is difficult to distinguish from different independent binding events, similar to Figure 2.16. For a more detailed discussion of cooperativity see Palmer and Bonner [11], Whitty [12], and Cornish-Bowden [13].

It should be realized that often full cooperativity is not realized, and the value of n derived from fitting a Hill model may not actually reflect the number of binding sites. The value of n may not even be an integer. The values should be interpreted as indicating that some cooperativity exists between the binding sites.

Key Concept: Cooperativity

In its broadest sense, cooperativity can be considered to include any set of molecular interactions in which the occurrence of some interactions affects the rate or affinity of other interactions, either via allosteric effects or configurational pre-organization.

Cooperativity can occur in monomeric enzymes, where the enzyme fluctuates between a high-affinity state (E_1) and a low-affinity state (E_2). To produce positive cooperativity, the equilibrium strongly favors the low-affinity conformation in the absence of substrate.

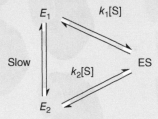

This is the simplest kinetic model with two unbound enzyme states that exhibit dynamic cooperativity.

More often, cooperativity is experienced in multi-site or multi-subunit enzymes, where binding of a molecule at one site influences the affinity for molecules binding at the other active sites. The affinity may be increased (positive cooperativity) or decreased (negative cooperativity). For more details on cooperativity see Palmer and Bonner [11], Whitty [12], and Cornish-Bowden [13].

2.8 Ligand Competition

Competition between ligands for binding to a macromolecule can often be used to compare affinities between ligands. This is because binding to a single binding site will be dependent on the relative affinities of the ligands for that site and the concentrations present. This is illustrated below, for different ligands, A and L having different affinities. The affinity for ligand L (green) is 3-fold higher than the affinity for ligand A (lilac), which means that the ML complex will be three times more abundant than the MA complex at equilibrium (Figure 2.18).

In this case, where ligand competition occurs, it is possible to calculate the concentration of one ligand-bound species, from a knowledge of the binding affinities and concentrations of each ligand as well as and the total concentration of protein. The equation describing the concentration of the ML complex, assuming tight-binding does not occur, is Equation 2.26.

$$[ML] = \frac{[M]_t[L]}{[L] + K_{dL}\left(1 + \frac{[A]}{K_{dA}}\right)}$$

Equation 2.26   Expression for binding of ligand, L, at a fixed concentration of competing ligand, A.

If [L] is varied with fixed [A], then plots will resemble Figure 2.19.
If [A] is varied with fixed [L], then plots will resemble Figure 2.20.

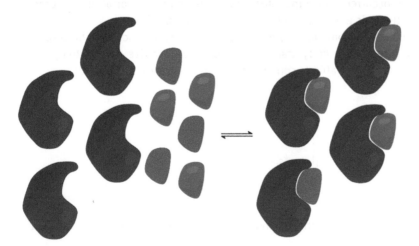

Figure 2.18   Enzyme binding to multiple ligands.
Competition between ligands for the same binding site. Binding of one ligand is mutually exclusive from binding of the other. The relative affinities for each ligand dictate the concentration of the species at equilibrium. Here the binding of ligand L (green) is 3-fold tighter than the binding of ligand A (lilac).

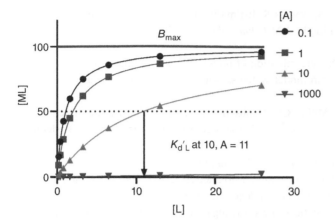

Figure 2.19 Ligand competition where the ligand L is varied.
Here, the effect of varying [L] is shown for different fixed concentrations of a competing ligand A. When [A] is low relative to its K_d, there is little competition. When [A] is high relative to its K_d, competition is greater, and less L binds. The apparent K_d ($K_d'{}_L$) is increased by the presence of increasing [A], as shown.

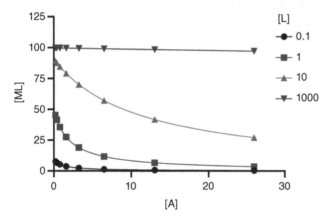

Figure 2.20 Ligand competition where the ligand A is varied.
Increasing concentration of ligand A reduces the concentration of ligand L bound to the protein, as A replaces L. However, if [L] is high relative to its K_d, then little competition is observed.

References

1 Voit EO, Martens HA, Omholt SW. 150 years of the mass action law. *PLoS Computational Biology.* 2015;11(1):e1004012.

2 Marcel Ovidiu V, Birgitt S. Mass action law versus local contagion dynamics. A mean-field statistical approach with application to the theory of epidemics. *Journal of Physics A: Mathematical and General.* 1996;29(16):4895.

3 Waage P, Gulberg CM. Studies concerning affinity. *Journal of Chemical Education.* 1986;63(12):1044.

4 Turnbull AP, Emsley P. Studying protein-ligand interactions using X-ray crystallography. *Methods in Molecular Biology.* 2013;1008:457–77.

5 Shortridge MD, Hage DS, Harbison GS, Powers R. Estimating protein-ligand binding affinity using high-throughput screening by NMR. *Journal of Combinatorial Chemistry.* 2008;10(6):948–58.

6 Fielding L, Rutherford S, Fletcher D. Determination of protein-ligand binding affinity by NMR: observations from serum albumin model systems. *Magnetic Resonance in Chemistry.* 2005;43(6):463–70.

7 Gao ZG, Chen A, Barak D, Kim SK, Müller CE, Jacobson KA. Identification by site-directed mutagenesis of residues involved in ligand recognition and activation of the human A3 adenosine receptor. *Journal of Biological Chemistry.* 2002;277(21):19056–63.

8 Edwards GB, Muthurajan UM, Bowerman S, Luger K. Analytical ultracentrifugation (AUC): an overview of the application of fluorescence and absorbance AUC to the study of biological macromolecules. *Current Protocols in Molecular Biology.* 2020;133(1):e131.

9 Yang TC, Catalano CE, Maluf NK. Analytical ultracentrifugation as a tool to study nonspecific protein-DNA interactions. *Methods in Enzymology.* 2015;562:305–30.

10 Ishii K, Noda M, Uchiyama S. Mass spectrometric analysis of protein-ligand interactions. *Biophysics and Physicobiology.* 2016;13:87–95.

11 Palmer T, Bonner PL. 13 – Sigmoidal Kinetics and Allosteric Enzymes. In: Palmer T, Bonner PL, editors. *Enzymes* (Second Edition): Woodhead Publishing; 2011. p. 239–54.

12 Whitty A. Cooperativity and biological complexity. *Nature Chemical Biology.* 2008;4(8):435–9.

13 Cornish-Bowden A. Understanding allosteric and cooperative interactions in enzymes. *The FEBS Journal.* 2014;281(2):621–32.

3

Protein QC and Handling

3.1   Introduction

Enzyme modulator screens and subsequent kinetic studies in early drug discovery are usually initially carried out on isolated recombinant proteins, following their expression and purification. Understanding the quality and behavior relative to the behavior expected for the protein in its more physiological cellular setting is critical. The protein reagent is present in every well of a high-throughput screen versus test compounds and so the integrity of the screen is completely dependent on the protein having the desired characteristics in order to identify modulators of its activity. Arguably, this is even more important than the quality control of any individual compound.

However, in the fast-paced setting of drug discovery, unfortunately, rigorous quality control of recombinant proteins sometimes may be overlooked or lack the thoroughness required. This may lead to irreproducible or even misleading results, which may have significant consequences for drug discovery projects in terms of both time and money wasted.

This chapter highlights the quality categories that should be interrogated, suggests techniques that may be used to assess protein quality in each category, and recommends a workflow for undertaking such studies.

3.2   Confirming Protein Identity

Clearly, before experiments are undertaken to characterize protein function or to identify or characterize small molecule modulators of activity, the identity of the protein should be verified. This involves confirming that the correct amino acid sequence has been expressed and that this sequence remains intact in the final enzyme preparation. This is important as often heterologous expression is used for protein production. This approach involves introducing complementary DNA encoding for the human protein of interest into the cell of another species so that the hosts' cellular

Laboratory Guide to Enzymology, First Edition. Geoffrey A. Holdgate, Antonia Turberville, and Alice Lanne.
© 2024 John Wiley & Sons, Inc. Published 2024 by John Wiley & Sons, Inc.

machinery will express the desired protein. This type of expression may be associated with missense substitutions, processivity errors, proteolyzed forms, or denatured or aggregated protein occurring at higher frequencies than for normal translation.

Several methods are available that provide confidence or proof that the protein that has been expressed is the one desired. For example, the identity may be checked using mass spectrometry (MS) techniques to determine intact protein mass, utilizing peptide mass fingerprinting or Edman sequencing.

3.2.1 Intact Mass Measurement by Mass Spectrometry (MS)

MS is a widely available method that can identify proteins according to their mass : charge ratio.

Firstly, a total ion chromatogram (TIC) is generated, this is the summed intensities of all the MS peaks against time. A spectrum shows the number of different ions detected in the sample distinguished by their mass : charge ratio and their intensity. The spectrum is used to reconstruct the intact mass spectrum peaks, give intact mass (in Da), and intensity for the proteins in the sample (Figure 3.1). If there are multiple proteins, then different intact mass spectrum peaks may be detected with different intensities at different mass : charge ratios. If the peak shows good agreement with the expected molecular mass for the protein, then confidence is provided that the purified protein is the expected one. Discrepancies signify a potential problem and highlight that further experiments are required to understand this difference [1–3]. A standard MS protocol can be found below in Protocol 3.1.

3.2.2 Peptide Mapping

MS detection may also be used in conjunction with proteolytic digests (often trypsin is used to produce peptides from the target protein) to confirm the identity of the protein. Once digested,

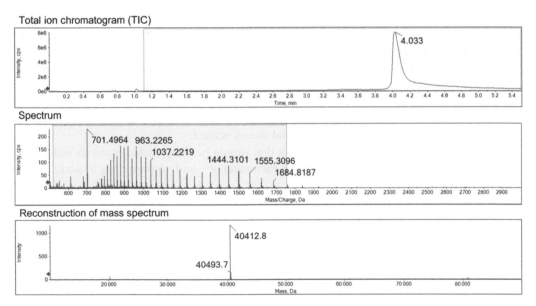

Figure 3.1 Intact mass chromatograms.
Total ion chromatogram is the summed intensity of all mass peaks in the sample. Spectrum is the different ions in the protein sample and their intensities. The spectrum is then reconstructed to give the intact mass of the proteins in the sample.

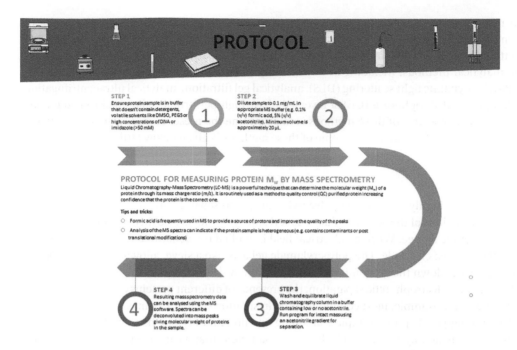

Protocol 3.1   Measuring intact protein mass by LC-MS.

the molecular weight of the resulting peptides is acquired using a high-mass accuracy instrument. These measured masses are compared with those from a database of masses expected for the protein and digestion method used. Protein identity is confirmed if several of the measured experimental masses match those predicted [4–6].

3.2.3   Edman Sequencing

In addition to MS methods, N-terminal Edman sequencing also can be used to confirm protein identity. Following tryptic digestion and separation, N-terminal Edman sequencing is performed on the peptides using a protein sequencer. In this process, the N-terminal amino group is derivatized, cleaved, and identified, with the procedure able to be repeated to identify the next amino acid. Digested peptides of up to around 30 amino acids can reliably be sequenced with the method. Many institutions have laboratories where these approaches may be accessed, or there are routes for obtaining these experiments commercially. Obtaining and interpreting sequencing data are important, as it allows the nature of the protein that is to be used in all ensuing work to be understood at this early stage, thus preventing issues that may arise with working with the wrong protein sequence subsequently [7–9].

3.3   Protein Purity

Once the protein's identity has been verified, it is important to understand how much of the correct sequence is contained within the total protein. This is important as the protein preparation may contain degradation products or host proteins that co-purify with the target enzyme (and which may have similar mass and other physical properties similar to the desired protein). In some instances, impurities may have activities similar to the target protein and these will need to be identified and removed before further work is undertaken. Purity may be assessed using a number of analytical methods, including sodium dodecyl sulfate-polyacrylamide gel electrophoresis (SDS-PAGE), dynamic light scattering (DLS), analytical gel filtration, analytical ultracentrifugation (AUC), capillary electrophoresis (CE) or reversed-phase high-performance liquid chromatography (RP-HPLC). The objective of these methods is to determine the presence and relative level of any contaminants. A brief explanation of some of these methods is given below [10].

3.3.1 SDS-PAGE

SDS-PAGE uses a highly cross-linked polyacrylamide gel to separate proteins by size. Under the denaturing conditions of the added sodium dodecyl sulfate, proteins lose their structure and adopt a uniform negative charge. When an electrical field is applied the denatured proteins will migrate towards the positive electrode. The polyacrylamide gel acts like a sieve, allowing smaller proteins to migrate further down the gel compared to larger proteins [11–13].

Following their electrophoretic separation, the presence of different proteins can be visualized by other methods, for example, most commonly protein staining such as Coomassie staining [14, 15] or silver staining [16, 17]. In a sample containing one protein only one band will be observed, a protein ladder is generally run alongside the protein sample so that the molecular weight of the pure protein can be estimated and indicate whether the pure protein is the desired protein (Figure 3.2). A standard SDS-PAGE protocol is shown below (Protocol 3.2).

3.3.2 Dynamic Light Scattering (DLS)

By shining laser light at large wavelengths (around 660 nm) on small particles light scattering occurs in all directions (so-called Rayleigh scattering). The intensity of the scattered light fluctuates over time due to the small particles in solution undergoing Brownian motion, with the distance between these scatterers constantly changing. By measuring how rapidly the intensity

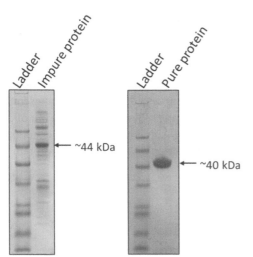

Figure 3.2 SDS PAGE.
Examples of protein samples after purification. The SDS-PAGE gel on the left shows an impure protein sample after purification, the desired protein (~44 kDa) is present in the protein sample but there are other proteins present in the sample. The SDS-PAGE gel on the right shows a pure protein sample after purification, the desired protein (~40 kDa) is the only protein present in the protein sample. Both SDS-PAGE gels have ladders that contain proteins of known molecular weight so the molecular weight of proteins present in the purified protein samples can be estimated.

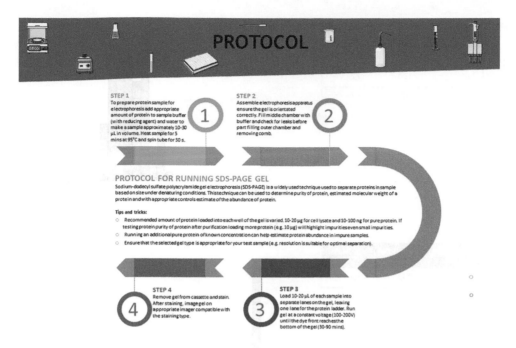

Protocol 3.2   Assessing protein purity using SDS-PAGE.

of the scattered light changes over time, information on the size of the particles can be obtained. In essence, the light scattering intensity of small particles changes more quickly over time than that of large particles. From observing the correlation between the scattered intensity of the initial and final states over time, a correlation function can be determined from which the diffusion coefficient, an average size, and a size distribution can be calculated. The particle diameter can then be calculated from the diffusion coefficient. The standard deviation of the diffusion coefficient informs on the width of the size distribution and gives a measure of the dispersity of the sample (Figure 3.3). The calculated polydispersity index (PI) is a measure of the heterogeneity of the sample based on size [18, 19].

3.3.3   Analytical Gel Filtration

Analytical gel filtration or size exclusion chromatography may be used to separate protein components in a mixture based on the molecular size and the hydrodynamic volume of those components. The separation occurs by differential exclusion or inclusion as the different protein components move through the stationary phase of the gel. Upon addition to the gel-containing column, proteins larger than the pore size are unable to diffuse into the beads, and so elute first. Smaller proteins or other components, ranging in size between very large and very small can enter the pores to varying degrees, governed by their size. Components smaller than the pores of the gel are able to enter the total pore volume and will be eluted last. The composition of the mobile phase is maintained during elution (isocratic elution) allowing conditions to be chosen that are compatible with the protein sample. To estimate the molecular weight of the test protein, the column must be first calibrated

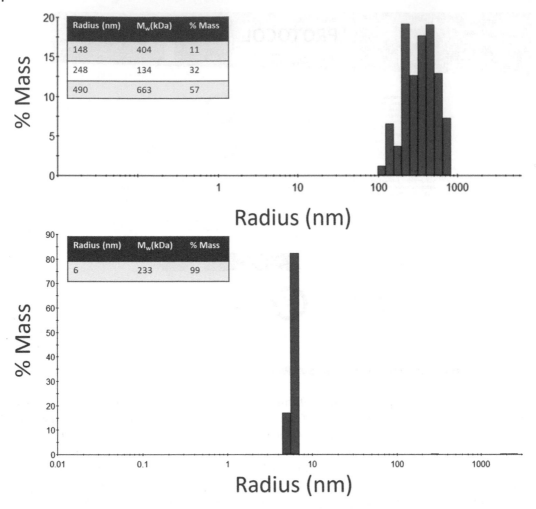

Figure 3.3 Dynamic light scattering (DLS).
Examples of protein samples tested in DLS. The top spectrum shows a polydisperse sample, the protein is in multiple different forms (e.g. aggregated, monomer, dimer), this is not a good quality protein and should not be used in drug discovery projects. The bottom spectrum shows a largely monodisperse sample (99.3%), this shows the protein is mainly present in one form, not aggregated, and of high quality.

with known protein standards (Figure 3.4). These standards can be bought commercially and span a wide range of molecular weights. Each standard's molecular weight will correspond to an elution volume and the molecular weight of the test protein can be estimated from this. For a pure protein, a symmetrical peak that elutes at the expected molecular weight indicates that the correct protein is observed. The presence of any additional peaks may indicate the presence of aggregates (peak before protein of interest) or degradation products (peaks after protein of interest) [20, 21].

3.3.4 Analytical Ultracentrifugation

AUC uses optical monitoring to determine the sedimentation profiles of components in a sample as they move under the influence of a strong gravitational force. Proteins have a different density from the solvent and so will sink or float in a strong enough field. Two approaches:

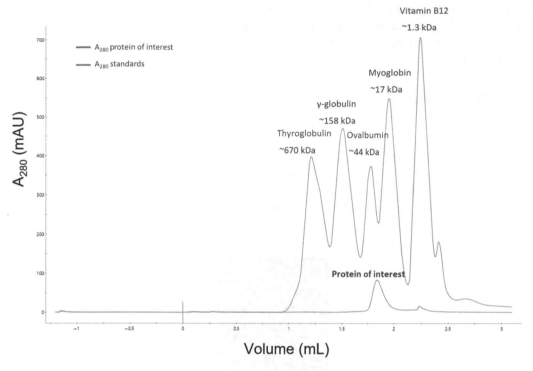

Figure 3.4 Analytical gel filtration.
Analytical gel filtration (or analytical size exclusion chromatography) allows the determination of the M_w of a protein. Commercial protein standards of known molecular weight are passed through the column. These protein standards will elute at specific elution volumes (the larger proteins first) and plotting of elution volume (accounting for the dead volume) against M_w allows calibration of the column. When the protein of interest is passed through the column the M_w of the protein can be estimated from the calibration.

sedimentation velocity (using observations of dynamic behavior) and sedimentation equilibrium (using observations in equilibrium) can provide information about protein size. However, sedimentation velocity experiments are mainly used to examine whether a protein sample is homogeneous in solution or a mixture of different components. Sedimentation equilibrium is typically employed for the determination of molecular weights. The presence of a "tail" at lower sedimentation coefficient values in a sedimentation velocity experiment often is indicative of heterogeneity as a result of contaminant proteins or degradation products of the protein of interest [22–24].

3.4 Concentration

Before a protein can be used for kinetic characterization or employed to identify modulators, the total concentration of protein should be measured. Although this is required, it is not sufficient as the total concentration may not represent the functional component (see below). However, the measurement of total protein concentration will provide an upper limit to the functional concentration of protein present. Knowledge of the protein concentration is required to ensure that suitable concentrations are used in subsequent experiments, for example, where enzyme concentration is required to be much lower than substrate or inhibitor concentrations, to avoid ligand depletion by binding and to allow the calculation of kinetic constants (e.g. from equations such as $k_{cat} = V_{max}/[E]_t$, see Chapter 5 for more detail). There are several methods to measure protein concentration that vary in their accuracy, sensitivity, and ease of use. Some of these methods are described below.

3.4.1 UV–Vis Spectrum

The concentration of a protein may be estimated spectrophotometrically from its absorbance at 280 nm and its calculated molar extinction coefficient. The absorption of radiation in the near UV region is determined by the amino acid content, mainly amino acids with aromatic rings such as tyrosine, tryptophan, and to a smaller extent phenylalanine. The presence of disulfide bonds, if present, also contributes to a minor degree. The molar extinction coefficient is calculated from the sequence and may be estimated using computer algorithms available from several sources, for example, web tools such as the ExPASy protein parameter tool (https://web.expasy.org/protparam).

Whilst measurement of concentration can be achieved by monitoring absorbance at 280 nm, further information can be obtained from a wavelength scan between 200 and 350 nm. In particular, nucleic acid contaminants may be identified as humps in the spectrum at 260 nm, resulting in a higher 260/280 nm absorbance ratio than expected (which are expected to be close to 0.57 for a pure protein sample) [25]. Monitoring the absorbance above 320 nm, where protein samples do not absorb light, allows the detection of large aggregates, where the signal generated is due to the scattering of light. A log–log plot of absorbance versus wavelength will generate a linear response that can be extrapolated back to 280 nm, the resulting antilog of which can be used to correct the measured protein concentration [26]. A standardized protocol for measuring protein concentration by UV–Vis spectroscopy is shown below (Protocol 3.3).

3.4.2 Bradford Assay

The principle behind the Bradford assay is the binding of Coomassie Brilliant Blue G-250 to protein carboxyl, and amino groups by van der Waals and electrostatic interactions, respectively. Under acidic conditions, the bound dye is converted to its anionic, blue form. When the dye is not bound to protein, the solution remains brown. This change in absorbance maximum for an acidic solution of Coomassie brilliant blue G-250 from 465 to 595 nm when bound to a protein occurs with an extinction coefficient that is relatively constant over at least a 10-fold concentration range. Applying Beer's law allows accurate quantitation of protein concentration by adding an appropriate volume of dye to allow measurement of the expected protein sample concentration [27, 28]. A protocol for a Bradford assay is shown below (Protocol 3.4).

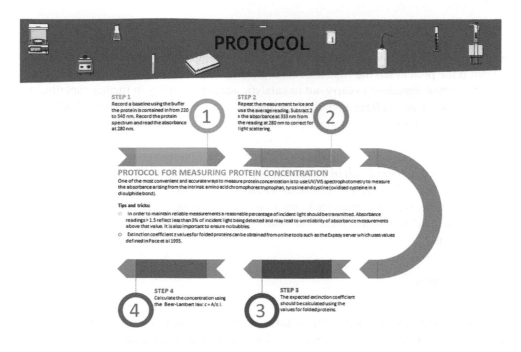

Protocol 3.3   Measuring protein concentration using UV–Vis spectroscopy.

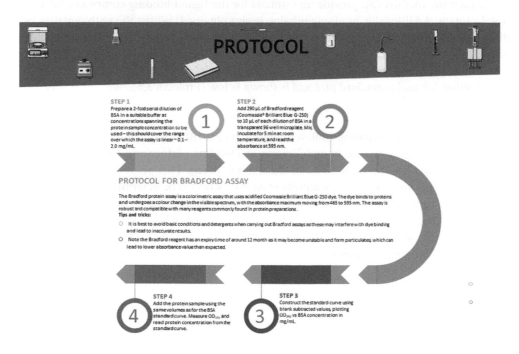

Protocol 3.4   Measuring protein concentration using a Bradford assay.

3.5 Functionality

Perhaps the most important quality control check is to determine that the enzyme has the desired activity. Even if the protein has the right sequence, is pure, and is at a useful concentration, without the ability to bind ligands or to carry out its catalytic activity, its utility in further experiments is limited. There are range of ligand binding approaches that could be employed to check those known ligands (tool compounds, substrates, or substrate analogs) bind with the expected affinity and mechanism. However, two that offer relatively straightforward setup and offer additional information are isothermal titration calorimetry (ITC) and surface plasmon resonance (SPR). These methods are described in more detail in Chapter 2, Section 2.5, but an overview of the desired outcome is provided below. Although proteins demonstrating the correct binding affinity for ligands but lacking catalytic activity (e.g. truncated protein domains) can be successfully used for ligand discovery, it is beneficial to establish functional assays in the early drug discovery cascade to assess inhibitory behavior.

3.5.1 Ligand Binding

3.5.1.1 ITC
ITC usually involves titrating a ligand solution into a protein sample and measuring the heat change that occurs as a result of the binding interaction. This technique allows measurement not only of the affinity of the interaction but also provides information on the stoichiometry of the binding event. Alternatively, if the total protein concentration and the binding stoichiometry for the ligand are known, then the method can provide an estimate for the ligand-binding competent fraction of the total protein. Additionally, binding enthalpy is also obtained. Whilst the enthalpy may be more difficult to interpret in terms of authentic function, its comparison across different batches of protein can be a useful metric to highlight potential issues. The method is usually rapid to perform, typically requiring an hour to complete a binding experiment. More details are provided in Chapter 2, Section 2.5, and a standard protocol is shown below (Protocol 3.5).

3.5.1.2 SPR
SPR and similar methods (including biolayer interferometry [BLI], and resonance waveguide grating [RWG] technologies) involve immobilization of a binding partner and monitoring optical changes that are related to the occurrence of the binding event. This allows a qualitative and quantitative measurement of the binding affinity of an interaction to be determined. If the protein can be used as the analyte (the molecule injected over the immobilized partner), in inhibition in solution assays [30], then it is possible to use this type of approach to also assess the protein concentration. Further details are provided in Chapter 2, Section 2.5. Protocol 3.6 describes a standard SPR ligand binding experiment.

3.5.2 Functional Studies

Ultimately, the only way to ensure the correct catalytic functionality of an enzyme preparation is to demonstrate that it can catalyze the desired reaction. This involves establishing an assay system to enable the detection of turnover of substrates to products and that this turnover requires or is enhanced by the presence of the enzyme. This then raises the question of how the assay should be configured. Ideally, a selective substrate should be used to demonstrate that the enzyme can specifically catalyze the turnover reaction. Lack of activity versus a selective substrate can be used

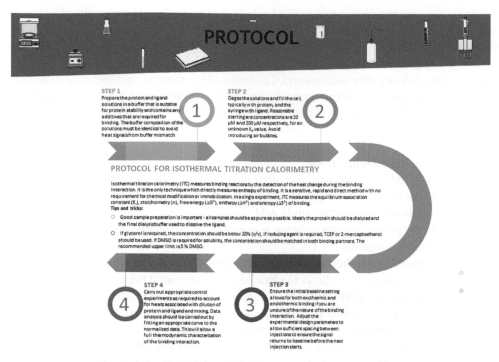

Protocol 3.5 Measuring ligand binding using ITC.
Source: Adapted from Cooper and Johnson [29].

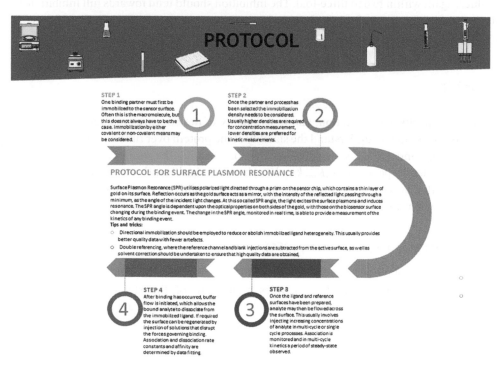

Protocol 3.6 Measuring ligand binding using SPR.
Source: Adapted from Myszka [31].

to illustrate that the enzyme preparation lacks the desired activity. The substrate with the lowest K_m (Michaelis–Menten constant, see Chapter 5) should be used preferentially, as this will require lower amounts of substrate to be used and should enable the activity of the enzyme with the highest activity in turning over the substrate to be detected. Use of less selective substrates may be problematic, as they can be potentially converted to product by a range of enzymes, and so do not offer the ability to demonstrate that the correct functionality is present, as this may result from an impurity contained in the enzyme preparation. For example, the substrate poly (Glu-Ala-Tyr) or the alternative poly (Glu-Tyr) may be phosphorylated by many tyrosine kinases and so may provide misleading information about the activity of the enzyme, especially if there is a small amount of highly functional contaminating kinase.

Ideally, low (around nM or below) concentrations of enzyme will suggest that enzyme is functioning as expected. The necessity for higher enzyme concentrations should raise questions over its functionality and its suitability for further study. Additionally, high concentrations of enzymes may increase the risk of contaminating activity being responsible for the observed turnover, especially when using less selective substrates.

Once the assay is established, it should be used to compare different batches of the enzyme preparation made at different times. This allows the batches to be compared between themselves and to other sources of the enzyme preparation, for example, those prepared commercially. Different enzyme batches may be compared by performing substrate dependence studies using the same substrate for each batch of enzyme. Usually, the parameter values, K_m and k_{cat}, using the same concentration of enzyme, should be within around two or three-fold of each other when the same enzyme activity is being compared (more detail in Chapter 5). Additionally, if an inhibitor is known, comparison of IC_{50} values (covered in Chapter 6) under the same assay conditions should yield similar values, again within two to three-fold. The inhibition should tend towards full inhibition at concentrations of inhibitor significantly above its IC_{50}, with more complex curves indicating issues with functionality or purity.

3.6 Stability

Once the identity, purity, concentration, and functionality of the enzyme preparation have been verified, it is also important to understand the stability of the protein over time and under different conditions of storage and use. This may take the form of repeating the studies above periodically over time to check that the parameters (K_m, k_{cat}, IC_{50}) have not significantly changed. Alternatively, there are methods that provide a route to measuring thermal stability, to verify that the same folding signature can be seen or to allow stability during catalysis to be determined.

3.6.1 Differential Scanning Calorimetry (DSC)

A macromolecule in solution exists in equilibrium between its folded and unfolded states. The thermal unfolding midpoint (T_m) provides a measure of the stability of the macromolecule, with a higher T_m, indicating a more stable macromolecule (Figure 3.5). In addition to the melting temperature, DSC also measures the enthalpy (ΔH) of unfolding that results from the heat-induced unfolding. The enthalpy of protein unfolding is estimated from the area under

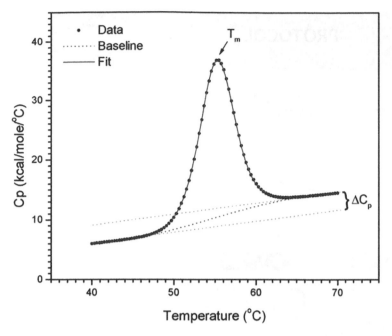

Figure 3.5 **Differential scanning calorimetry (DSC) thermogram.**
The DSC thermogram shows the change in heat capacity, C_p, as the protein unfolds with temperature. The change in heat capacity between the folded and unfolded states is ΔC_p. The mid-point of the unfolding transition is the melting temperature, T_m. The area under the curve provides the enthalpy of unfolding. Note the baseline during the unfolding transition must be extrapolated between the pre- and post-transition baselines.

the concentration-normalized DSC peak. ΔH represents the total energy taken up in raising the system to temperature T at constant pressure. For a macromolecule, this describes the energy (heat) taken to unfold the macromolecule. The ΔH for unfolding is positive and so describes an endothermic process [32, 33].

The relative ΔH values from the different thermograms may be used to estimate the amount of folded macromolecule in each sample when compared to a reference thermogram with 100% folded macromolecule. From the difference in the pre- and post-transition baselines, the change in heat capacity (ΔC_p) may also be measured. Using DSC, it is possible to evaluate the thermal stability as a measure of initial stability and to compare the stability of long-term storage. Stable proteins will likely have fewer issues during use, are more cost-effective to produce, and will have a better chance of remaining functional during storage with fewer issues such as aggregation. A standard protocol for this can be found below (Protocol 3.7) [34].

3.6.2 Differential Scanning Fluorimetry (DSF)

Differential scanning fluorimetry (DSF) measures protein unfolding by monitoring changes in fluorescence as a function of temperature. The conventional DSF approach employs a hydrophobic dye that binds to exposed hydrophobic patches on proteins as they are exposed upon unfolding (Figure 3.6). The method uses a real-time PCR instrument to monitor the change in fluorescence

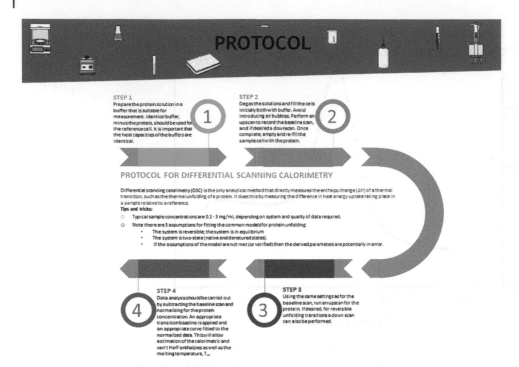

PROTOCOL

STEP 1
Prepare the protein solution in a buffer that is suitable for measurement. Identical buffer, minus the protein, should be used for the reference cell. It is important that the heat capacities of the buffers are identical.

STEP 2
Degas the solutions and fill the cells initially both with buffer. Avoid introducing air bubbles. Perform an upscan to record the baseline scan, and if desired a downscan. Once complete, empty and re-fill the sample cell with the protein.

PROTOCOL FOR DIFFERENTIAL SCANNING CALORIMETRY

Differential scanning calorimetry (DSC) is the only analytical method that directly measures the enthalpy change (ΔH) of a thermal transition, such as the thermal unfolding of a protein. It does this by measuring the difference in heat energy uptake taking place in a sample relative to a reference.
Tips and tricks:
○ Typical sample concentrations are 0.1 - 3 mg/ml, depending on system and quality of data required.
○ Note there are 3 assumptions for fitting the common models for protein unfolding:
• The system is reversible; the system is in equilibrium
• The system is two-state (native and denatured states).
• If the assumptions of the model are not met (or verified) then the derived parameters are potentially in error.

STEP 4
Data analysis should be carried out by subtracting the baseline scan and normalizing for the protein concentration. An appropriate transition baseline is applied and an appropriate curve fitted to the normalized data. This will allow estimation of the calorimetric and van't Hoff enthalpies as well as the melting temperature, T_m.

STEP 3
Using the same settings as for the baseline scan, run an upscan for the protein. If desired, for reversible unfolding transitions a down scan can also be performed.

Protocol 3.7   Measuring protein stability using differential scanning calorimetry (DSC).

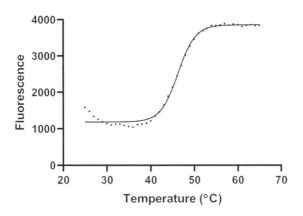

Figure 3.6   A Differential scanning fluorimetry (DSF) plot.
The DSF trace shows the change in fluorescence of dye as it binds to exposed hydrophobic regions of the protein as it unfolds. The data above are fitted to a simple Boltzmann expression to generate fluorescence values for the folded and unfolded states. The point of inflection provides the T_m value, and the slope of the curve is also fitted.

occurring as the dye binds to the unfolded protein and its fluorescence yield increases as its environment changes from the polar solvent to the hydrophobic setting of the protein (see Figure 3.6). Although the method is indirect compared to DSC, relying on the binding of the dye to detect unfolding, it has higher throughput, consumes smaller amounts of protein, being able to be conducted in 96 or 384-well-plate formats and so is well suited to buffer compatibility studies or even ligand binding screens [35, 36]. Protocol 3.8 gives a standard protocol for DSF [37].

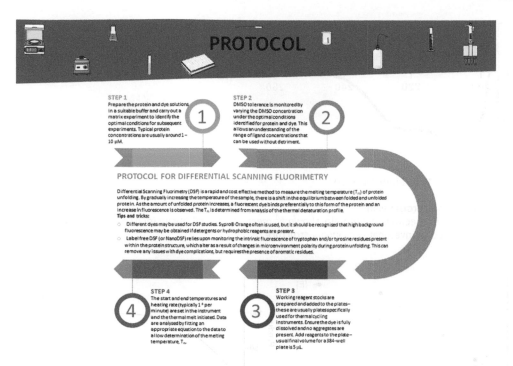

Protocol 3.8   Measuring protein stability using differential scanning fluorimetry (DSF). Adapted from [34].

3.6.3   Circular Dichroism

Circular dichroism (CD) spectroscopy is a well-established method for studying the secondary structure of proteins. As such, it also can be used to follow the stability, dynamics, folding, and interactions of proteins. The basis for the technique is the differential absorption of left- and right-handed circularly polarized light by molecules containing a chiral center or having a three-dimensional structure that provides a chiral environment. Optically active chiral molecules preferentially absorb one component of the circularly polarized light. Protein secondary structural information may be interpreted from the CD signals in the far ultraviolet (UV) wavelength region between 240 and 190 nm due to the amide chromophores of the peptide bonds. The common secondary structural elements: α-helices, β-sheets, β-turn, and random coil conformations all have characteristic spectra, which contribute to the protein secondary structure analysis (Figure 3.7). CD allows the exploration of a wide range of conditions and temperatures, with consumption of relatively small amounts of protein and moderately rapid data collection [38, 39].

3.6.4   Selwyn's Test

If within a series of enzyme assays, assay parameters such as the substrate concentration, any required activator concentration, and other factors like temperature are kept constant, the amount of product generated is a function of the enzyme concentration and time, where $[P] = f[E] \cdot t$. Observing the progress curve in this way, provides some advantages when the reaction rate decreases rapidly as the reaction progresses – a situation which may result from inactivation of the enzyme – making estimation of the initial rate difficult as well as inaccurate. Understanding if any enzyme-dependent inactivation is present should be accomplished before the results of any

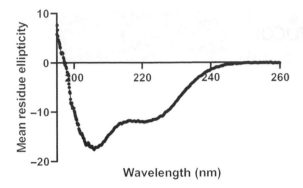

Figure 3.7 Circular dichroism scan.
The circular dichroism spectrum of a protein, shown as mean residue ellipticity (deg cm^2 dmol^{-1}) versus wavelength (nm) used to investigate secondary structure.

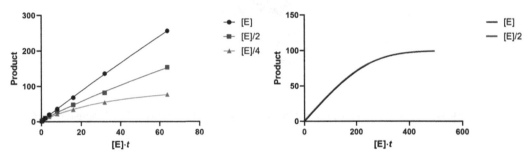

Figure 3.8 Selwyn's test.
Selwyn's test is the test for enzyme inactivation. When enzyme inactivation occurs, there is an observable decrease in the product when plotted versus [E]·t (product of enzyme concentration and time) and increased curvature as the enzyme concentration is halved, this is shown in the left panel. When there is no enzyme inactivation, there is no difference in product versus [E]·t plots as the enzyme concentration is changed, seen in the right panel.

subsequent assays can be interpreted. Selwyn's test distinguishes between causes of curvature in progress curves, such as substrate depletion or product inhibition, which can be analyzed using the appropriate kinetic analysis, and those which are not suitable for such analysis, such as enzyme denaturation. If the enzyme becomes denatured during the reaction, the concentration of active enzyme is a time-dependent quantity, and so the expression above no longer holds. In order to check for this situation, different concentrations of enzyme are utilized, plotting the amount of product generated versus the concentration of enzyme multiplied by time. If, for these sets of progress curves, all the points should fall on one line, this demonstrates that there is no time-dependent inactivation (see Figure 3.8). However, where the points for different concentrations of enzyme fall on different curves, this demonstrates a unimolecular denaturation event [40].

References

1 Heck AJR, van den Heuvel RHH. Investigation of intact protein complexes by mass spectrometry. *Mass Spectrometry Reviews*. 2004;23(5):368–89.

2 Donnelly DP, Rawlins CM, DeHart CJ, Fornelli L, Schachner LF, Lin Z, et al. Best practices and benchmarks for intact protein analysis for top-down mass spectrometry. *Nature Methods.* 2019;16(7):587–94.

3 Tipton JD, Tran JC, Catherman AD, Ahlf DR, Durbin KR, Kelleher NL. Analysis of intact protein isoforms by mass spectrometry. *Journal of Biological Chemistry.* 2011;286(29):25451–8.

4 Zhang W, Chait BT. ProFound: an expert system for protein identification using mass spectrometric peptide mapping information. *Analytical Chemistry.* 2000;72(11):2482–9.

5 Egelhofer V, Gobom J, Seitz H, Giavalisco P, Lehrach H, Nordhoff E. Protein identification by MALDI-TOF-MS peptide mapping: a new strategy. *Analytical Chemistry.* 2002;74(8):1760–71.

6 Judd RC. Comparison of protein primary structures. Peptide mapping. *Methods in Molecular Biology (Clifton, NJ).* 1994;32:185–205.

7 Williams N, Pineda F, Lam TT, Bruce C, Bingham J, Hodsdon M, et al. Edman sequencing and amino acid analysis in the proteomic age. *The FASEB Journal.* 2013;27(S1):790.11–11.

8 Hunkapiller MW, Hood LE. Protein sequence analysis: automated microsequencing. *Science.* 1983;219(4585):650–9.

9 Brune DC, Hampton B, Kobayashi R, Leone JW, Linse KD, Pohl J, et al. ABRF ESRG 2006 study: Edman sequencing as a method for polypeptide quantitation. *Journal of Biomolecular Techniques: JBT.* 2007;18(5):306–20.

10 Rhodes DG, Laue TM. 38 – Determination of Protein Purity. In: Burgess RR, Deutscher MP, editors. *Methods in Enzymology.* 463: Academic Press; 2009. p. 677–89.

11 Gallagher SR. SDS-polyacrylamide gel electrophoresis (SDS-PAGE). *Current Protocols Essential Laboratory Techniques.* 2012;6(1):7.3.1–28.

12 Manns JM. SDS-polyacrylamide gel electrophoresis (SDS-PAGE) of proteins. *Current Protocols in Microbiology.* 2011;22(1):A.3M.1–13.

13 Mohan SB. Determination of Purity and Yield. In: Kenney A, Fowell S, editors. *Practical Protein Chromatography.* Totowa, NJ: Humana Press; 1992. p. 307–23.

14 Brunelle JL, Green R. 13 – Coomassie blue staining. In: Lorsch J, editor. *Methods in Enzymology.* 541: Academic Press; 2014. p. 161–7.

15 Gauci VJ, Padula MP, Coorssen JR. Coomassie blue staining for high sensitivity gel-based proteomics. *Journal of Proteomics.* 2013;90:96–106.

16 Chevallet M, Luche S, Rabilloud T. Silver staining of proteins in polyacrylamide gels. *Nature Protocols.* 2006;1(4):1852–8.

17 Wray W, Boulikas T, Wray VP, Hancock R. Silver staining of proteins in polyacrylamide gels. *Analytical Biochemistry.* 1981;118(1):197–203.

18 Lorber B, Fischer F, Bailly M, Roy H, Kern D. Protein analysis by dynamic light scattering: Methods and techniques for students. *Biochemistry and Molecular Biology Education.* 2012;40(6):372–82.

19 Stetefeld J, McKenna SA, Patel TR. Dynamic light scattering: a practical guide and applications in biomedical sciences. *Biophysical Reviews.* 2016;8(4):409–27.

20 Ó'Fágáin C, Cummins PM, O'Connor BF. Gel-filtration chromatography. *Methods in Molecular Biology (Clifton, NJ).* 2017;1485:15–25.

21 Irvine GB. Determination of molecular size by size-exclusion chromatography (gel filtration). *Current Protocols in Cell Biology.* 2000;6(1):5.1–5.16.

22 Cole JL, Lary JW, Moody TP, Laue TM. Analytical ultracentrifugation: sedimentation velocity and sedimentation equilibrium. *Methods in Cell Biology.* 2008;84:143–79.

23 Lebowitz J, Lewis MS, Schuck P. Modern analytical ultracentrifugation in protein science: a tutorial review. *Protein Science: A Publication of the Protein Society.* 2002;11(9):2067–79.

24 Edwards GB, Muthurajan UM, Bowerman S, Luger K. Analytical ultracentrifugation (AUC): an overview of the application of fluorescence and absorbance AUC to the study of biological macromolecules. *Current Protocols in Molecular Biology*. 2020;133(1):e131.

25 Glasel JA. Validity of nucleic acid purities monitored by 260 nm/280 nm absorbance ratios. *BioTechniques*. 1995;18(1):62–3.

26 Leach SJ, Scheraga HA. Effect of light scattering on ultraviolet difference spectra. *Journal of the American Chemical Society*. 1960;82(18):4790–2.

27 Brady PN, Macnaughtan MA. Evaluation of colorimetric assays for analyzing reductively methylated proteins: biases and mechanistic insights. *Analytical Biochemistry*. 2015;491:43–51.

28 Bradford MM. A rapid and sensitive method for the quantitation of microgram quantities of protein utilizing the principle of protein-dye binding. *Analytical Biochemistry*. 1976;72:248–54.

29 Cooper A, Johnson CM. Isothermal titration microcalorimetry. *Methods in Molecular Biology (Clifton, NJ)*. 1994;22:137–50.

30 de Mol NJ. Affinity constants for small molecules from SPR competition experiments. *Methods in Molecular Biology (Clifton, NJ)*. 2010;627:101–11.

31 Myszka DG. Improving biosensor analysis. *Journal of Molecular Recognition: JMR*. 1999;12(5):279–84.

32 Johnson CM. Differential scanning calorimetry as a tool for protein folding and stability. *Archives of Biochemistry and Biophysics*. 2013;531(1):100–9.

33 Bruylants G, Wouters J, Michaux C. Differential scanning calorimetry in life science: thermodynamics, stability, molecular recognition and application in drug design. *Current Medicinal Chemistry*. 2005;12(17):2011–20.

34 Cooper A, Johnson CM. Differential scanning calorimetry. *Methods in Molecular Biology (Clifton, NJ)*. 1994;22:125–36.

35 Gao K, Oerlemans R, Groves MR. Theory and applications of differential scanning fluorimetry in early-stage drug discovery. *Biophysical Reviews*. 2020;12(1):85–104.

36 Malik K, Matejtschuk P, Thelwell C, Burns CJ. Differential scanning fluorimetry: rapid screening of formulations that promote the stability of reference preparations. *Journal of Pharmaceutical and Biomedical Analysis*. 2013;77:163–6.

37 Alexander CG, Wanner R, Johnson CM, Breitsprecher D, Winter G, Duhr S, et al. Novel microscale approaches for easy, rapid determination of protein stability in academic and commercial settings. *Biochimica et Biophysica Acta*. 2014;1844(12):2241–50.

38 Greenfield NJ. Using circular dichroism spectra to estimate protein secondary structure. *Nature Protocols*. 2006;1(6):2876–90.

39 Haque MA, Kaur P, Islam A, Hassan MI. Application of circular dichroism spectroscopy in studying protein folding, stability, and interaction. *Advances in Protein Molecular and Structural Biology Methods*: Elsevier; 2022. p. 213–24.

40 Selwyn MJ. A simple test for inactivation of an enzyme during assay. *Biochimica et Biophysica Acta (BBA) - Enzymology and Biological Oxidation*. 1965;105(1):193–5.

4

Buffers and their Use in the Study of Enzyme Mechanism

4.1 Introduction

The activity of an enzyme is dependent upon the concentration of hydrogen ions, generally measured using the pH scale, present in the solution in which the enzyme activity is being studied. The concentration of hydrogen ions affects the ionization and hence the net charges on some of the constituent amino acids of the protein (and/or potentially the substrate), and so may alter the network of intramolecular forces that govern the structure and activity of the enzyme. Although some enzymes exhibit activity over a broad range of pH (e.g. amylase) [1, 2], many show maximal activity at a narrow pH range (e.g. pepsin) [3, 4]. Understanding pH, how to control it, and the effects it has on enzyme activity is a key consideration for designing suitable assays for enzyme kinetic studies.

4.1.1 Ionization and pK_a

Ionization is the complete loss of an electron from an atomic or molecular species. When this occurs, the resulting species is called an ion.

Key Concept: Ionization and Acid-base Characterization

Arrhenius's theory classifies compounds as acids or bases, depending on whether they dissociate in water to produce hydrogen (H^+) or hydroxide (OH^-) ions. Arrhenius acids dissociate in water to form hydrogen (H^+) ions, whereas Arrhenius bases dissociate in water to form hydroxide (OH^-) ions. There are alternative definitions of acids and bases. The Brønsted-Lowry

Laboratory Guide to Enzymology, First Edition. Geoffrey A. Holdgate, Antonia Turberville, and Alice Lanne.
© 2024 John Wiley & Sons, Inc. Published 2024 by John Wiley & Sons, Inc.

definition states that acids are proton (H^+) donors, while bases are proton (H^+) acceptors. Every Arrhenius acid or base is considered a Brønsted-Lowry acid or base.

A third definition is the Lewis definition, where a Lewis acid is considered an electron acceptor, and a Lewis base is considered an electron donor.

For example, a species, X may lose an electron, as shown in Reaction scheme 4.1.

$$X \longrightarrow X^+ + e^-$$

Reaction scheme 4.1   Reaction scheme for ionization of species X by the loss of an electron.
X^+ is the ion, and its positive charge means it is termed a cation.

Ions may further ionize to form higher charged species, see Reaction scheme 4.2.

$$X \longrightarrow X^+ + e^-$$
$$X^+ \longrightarrow X^{2+} + e^-$$
$$X^{2+} \longrightarrow X^{3+} + e^-$$

Reaction scheme 4.2   Reaction scheme for ionization of an ion to form a higher charged species.

Although strictly ionization refers to the loss of an electron to form a positive ion, it is also frequently used to describe the situation where a species gains an electron to form a negative ion (anion), see Reaction scheme 4.3.

$$X + e^- \longrightarrow X^-$$

Reaction scheme 4.3   Reaction scheme for ionization of species X by the gain of an electron.

The energy required to remove an electron from a species is its ionization energy. The ionization of molecules occurs readily in solution, for example, water molecules exist in an equilibrium with ions, formed from self-ionization (Reaction scheme 4.4).

$$2H_2O \rightleftharpoons H_3O^+ + OH^-$$

Reaction scheme 4.4   The ionization of water.

Consider the ionization of a monoprotic acid shown below (Reaction scheme 4.5).

$$HA_{(aq)} \xrightleftharpoons{K_a} A^-_{(aq)} + H^+_{(aq)}$$

Reaction scheme 4.5   Reaction scheme for the ionization of a monoprotic acid.

Where K_a is the ionization (or dissociation) constant for HA and is given by the expression (Equation 4.1)

$$K_a = \frac{[A^-][H^+]}{[HA]}$$

Equation 4.1   Expression for K_a.
The K_a represents the acid dissociation constant. The larger the K_a value, the stronger the acid.

Therefore, the equation for pK_a is shown in Equation 4.2.

$$pK_a = -\log K_a = \log \frac{[HA]}{[A^-][H^+]}$$

Equation 4.2   Expression for pK_a.
The pK_a represents the negative log of the acid dissociation constant (K_a). The smaller the value of pK_a, the more dissociation occurs and the stronger the acid.

4.1.2   pH

The pH value is a number that expresses the acidity or alkalinity of an aqueous solution by expressing the hydrogen ion concentration on a logarithmic scale (Equation 4.3).

$$pH = -\log[H^+]$$

Equation 4.3   Expression for pH.
The pH value expresses the hydrogen ion concentration on the logarithmic scale. [H$^+$] is measured in moles per liter.

The range of pH extends from 0 to 14, where solutions with a pH value of less than 7 are acidic ($[H^+] > [OH^-]$ or more precisely $[H_3O^+] > [OH^-]$), and solutions with a pH value greater than 7 are basic ($[H^+] < [OH^-]$). Solutions with a pH of 7 are neutral ($[H^+] = [OH^-]$), with measurements taking place at 25 °C. It should be remembered that temperature has a significant effect on pH measurements. For example, for pure water, as the temperature rises, molecular vibrations increase, increasing the ability of water to ionize and form more hydrogen ions. As a result, the pH will drop. For example, the pH of pure water at 25 °C is 7, whereas at 50 °C it is 6.63.

The pH is usually measured with a pH meter, which converts the difference in electromotive force (electrical potential or voltage) between electrodes placed in the solution to be measured, into the pH reading. Rough pH measurements may be made using litmus paper or other types of pH paper, which changes color at certain pH values. Although most indicators and pH papers identify pH within a narrow range, universal indicators can provide a color change over a wider range of pH values.

The International Union of Pure and Applied Chemistry (IUPAC) employs a slightly different pH (Equation 4.4) scale based on electrochemical measurements of a standard buffer solution.

$$pH = -\log\left[a_{H^+}\right]$$

Equation 4.4   Expression for the IUPAC definition of pH.
a_{H^+} represents the hydrogen activity, which is the effective concentration of hydrogen ions in a solution and may vary slightly from the true concentration.

4.1.3   Henderson-Hasselbalch Equation

From rearranging the expression for the dissociation of a monoprotic acid above (Reaction scheme 4.5) and the expression for pH shown in Equation 4.3, an expression relating pH and pK_a may be obtained (Equation 4.5). Using this equation, and a knowledge of the pK_a, it is possible to calculate the required ratio of weak acid and conjugate base required for a particular pH value.

$$pH = pK_a + \log \frac{[A^-]}{[HA]}$$

Equation 4.5   Expression for the Henderson-Hasselbalch equation.

When pH = pK_a, then $\log \frac{[A^-]}{[HA]} = 0$, and so $[A^-] = [HA]$ (because $\log_{10} 1 = 0$).

4.1.4   Buffers

A buffer solution is one that can avoid a change in pH when small amounts of acid or alkali are added. The buffer is able to neutralize the small amounts of added acid or base (in the form of H_3O^+ and OH^-) and is therefore able to maintain the pH of the solution over a narrow range of added ions. A buffer solution must contain substances that will remove any hydrogen ions or hydroxide ions that are added to it, to prevent the pH from changing.

To successfully maintain the pH of a solution, a buffer usually consists of a weak conjugate acid-base pair. Two situations are possible: a weak acid and its conjugate base, or a weak base and its conjugate acid. The choice of the appropriate combination will depend upon the chosen pH for the buffer solution.

An example is ethanoic acid (CH_3COOH) and a salt containing its conjugate base, the ethanoate anion (CH_3COO^-), for example, sodium ethanoate (CH_3COONa). Ethanoic acid is a weak acid and so the position of the equilibrium between ethanoic acid and the ethanoate ion in solution lies over to the left (Reaction scheme 4.6).

$$CH_3COOH_{(aq)} \rightleftharpoons CH_3COO^-_{(aq)} + H^+_{(aq)}$$

Reaction scheme 4.6   Reaction scheme for the ethanoic acid dissociation.
Addition of sodium ethanoate to this solution provides more ethanoate ions, and following Le Chatelier's Principle, this results in the position of the equilibrium moving even further to the left. The solution now contains a significant amount of non-ionized ethanoic acid, a significant amount of ethanoate ions, and enough hydrogen ions to make the final solution acidic.

Now consider the addition of hydrogen ions to the solution. To avoid a decrease in pH, the buffer must remove most of the added H^+. This occurs as the hydrogen ions combine with the ethanoate ions to form ethanoic acid as shown in Reaction scheme 4.7.

$$CH_3COO^-_{(aq)} + H^+_{(aq)} \rightleftharpoons CH_3COOH_{(aq)}$$

Reaction scheme 4.7   Reaction scheme for addition of hydrogen ions to a solution of ethanoic acid.

Conversely, upon addition of hydroxide ions, the buffer must remove these to avoid an increase in pH. There are two ways this can occur. The first is by reacting with the ethanoic acid (Reaction scheme 4.8).

$$CH_3COOH_{(aq)} + OH^-_{(aq)} \rightleftharpoons CH_3COO^-_{(aq)} + H_2O_{(l)}$$

Reaction scheme 4.8   Reaction scheme for addition of hydroxide ions to a solution of ethanoic acid – reaction with ethanoic acid.

The added hydroxide ions may also react with hydrogen ions present to form water. Since this removes hydrogen ions, the equilibrium moves to replace them (Reaction scheme 4.9), in this way, most of the added OH^- ions are removed.

$$CH_3COOH_{(aq)} \rightleftharpoons CH_3COO^-_{(aq)} + H^+_{(aq)}$$

Reaction scheme 4.9 Reaction scheme for the shift in the equilibrium position to replace hydrogen ions removed by reaction with added hydroxide ions.

The example above demonstrates the behavior of an acidic buffer solution. However, buffer solutions can also be alkaline. For example, consider the weak base ammonia (NH_3), and a salt of its conjugate acid, ammonium nitrate (NH_4NO_3). The position of the equilibrium lies over to the left (Reaction scheme 4.10).

$$NH_{3(aq)} + H_2O_{(l)} \rightleftharpoons NH_4^+{}_{(aq)} + OH^-_{(aq)}$$

Reaction scheme 4.10 Reaction scheme for the ammonia dissociation reaction.

Addition of ammonium nitrate increases the concentration of ammonium ions, and the equilibrium moves, according to Le Chatelier's principle, even more to the left. This time the solution contains a significant amount of non-ionized ammonia, a significant amount of ammonium ions, and enough hydroxide ions to make the solution alkaline.

When hydrogen ions are added to this solution, there are two ways the additional ions can be removed to avoid a drop in pH. Firstly, the hydrogen ions could react with ammonia to form an ammonium ion (Reaction scheme 4.11).

$$NH_{3(aq)} + H^+_{(aq)} \rightleftharpoons NH_4^+{}_{(aq)}$$

Reaction scheme 4.11 Reaction scheme for the addition of hydrogen ions to a solution of ammonia – reaction with ammonia.

Secondly, the H^+ ions could react with hydroxide ions to form water. Since this removes hydroxide ions, the equilibrium moves to replace them (Reaction scheme 4.12).

$$NH_{3(aq)} + H_2O_{(l)} \rightleftharpoons NH_4^+{}_{(aq)} + OH^-_{(aq)}$$

Reaction scheme 4.12 Reaction scheme for the shift in the equilibrium position to replace hydroxide ions removed by reaction with added hydrogen ions.

Upon addition of hydroxide ions, these are removed by reaction with ammonium ions (Reaction scheme 4.13).

$$NH_4^+{}_{(aq)} + OH^-_{(aq)} \rightleftharpoons NH_{3(aq)} + H_2O_{(l)}$$

Reaction scheme 4.13 Reaction scheme for the addition of hydroxide ions to a solution of ammonia.

Key Concept: Le Chatelier's Principle

This states that any change made in a system at equilibrium results in a change of the equilibrium in the direction that minimizes the change. In the original version from 1884, Le Chatelier referenced only pressure, but it was soon realized that the principle covers any kind of external influence.

4.2 Buffering Capacity

The buffering capacity describes the ability of a buffer solution to resist changes in pH by its ability to neutralize H^+ and OH^- ions. The effect on the pH can vary when an acid or base is added to the buffer depending on both the initial pH and the capacity of the buffer to resist the change in pH. The buffer capacity (β) is defined as the number of moles of acid or base (n) that must be added to 1 L of the buffer solution to increase or decrease the pH by one unit (Equation 4.6).

$$\beta = \frac{n}{\Delta pH}$$

Equation 4.6   Expression for buffering capacity.
β is the buffering capacity, n is number of moles of acid or base and ΔpH is the change in pH.

The higher the buffer capacity, the more acid or base can be added before the pH of the buffer solution changes significantly. Thus, the higher the concentration of the weak acid and its conjugate base (or a weak base and its conjugate acid) a buffer has, the higher its buffering capacity will be.

If a buffer has a greater concentration of its conjugate base than the weak acid, it will have a higher tolerance for added acid. Conversely, if a buffer has a greater concentration of the weak acid than that of its conjugate base, then it will have a higher resistance to added base.

Since the value of the buffer capacity is dependent upon the concentrations of the components and increases with their concentration, it can be easily understood that buffer solutions with a pH equal to the pK_a value of the acid (used to make this solution) have the greatest buffering capacity (Figure 4.1), since high concentrations of both conjugate acid and conjugate base are present, see the Henderson-Hasselbalch equation above (Equation 4.5).

4.3 Ionic Strength

The ionic strength (I) of a solution is a measure of the concentration of charges contained within the solution. The ionic strength is given by the following expression (Equation 4.7).

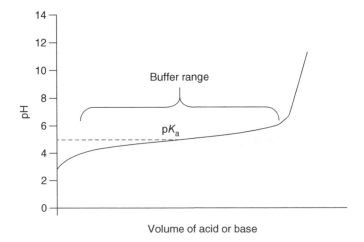

Figure 4.1   Buffering capacity.
The useful buffer range extends approximately within the range of pH values from 2 units below pK_a, to 2 units above, with the ideal range $pK_a \pm 1$ unit.

$$I = \frac{1}{2} \sum C_i z_i^2$$

Equation 4.7 Expression for ionic strength.
where C_i is the molar concentration of the ion and z_i is the net charge of the ion. Note that half is required because both anions and cations are included.

For electrolytes where each ion is singularly charged, the ionic strength is simply equal to the concentration. However, the situation becomes more complicated for multivalent ions, which contribute more strongly to the ionic strength. For example, an electrolyte where each species is a bivalent ion, the ionic strength is equal to four times the concentration. For mixed species, the calculation is more complex. For example, the ionic strength of a solution of 0.5 M potassium chloride ($KCl \rightarrow K^+ + Cl^-$) and 0.1 M magnesium sulfate ($MgSO_4 \rightarrow Mg^{2+} + SO_4^{2-}$) is calculated as follows (Calculation 4.1).

$$1 = \frac{1}{2}[0.5(+1)^2 + 0.5(-1)^2 + 0.1(+2)^2 + 0.1(-2)^2] = 0.9 \text{ M}$$

Calculation 4.1 Expression for the calculation of the ionic strength of potassium chloride.

Often, ionic strength in enzyme solutions is controlled or dominated by adding relatively large amounts of NaCl, where the ionic strength is equal to the concentration. However, it must be remembered that buffer components also contribute to the ionic strength of the solution.

Consider the use of a 50 mM ethanoic acid/sodium ethanoate buffer as described in Equation 4.7, used at a pH of 5. From the Henderson-Hasselbalch equation, we can calculate the relative concentrations of salt and acid, as shown in Calculation 4.2.

$$pH = pK_a + Log\frac{[A^-]}{[HA]}$$ Substitute in values

$$5 = 4.76 + Log\frac{[CH_3COO^-]}{[CH_3COOH]}$$ Subtract 4.76 from each side

$$0.24 = Log\frac{[CH_3COO^-]}{[CH_3COOH]}$$ Multiply to 10th power

$$1.74 = \frac{[CH_3COO^-]}{[CH_3COOH]}$$ Multiply by $[CH_3COOH]$ on both sides

$$1.74[CH_3COOH] = [CH_3COO^-]$$

$$[CH_3COOH] + [CH_3COO^-] = 50 \text{ mM}$$ Substitute $[CH_3COO^-]$ for $1.74[CH_3COOH]$

$$2.74[CH_3COOH] = 50 \text{ mM}$$ Divide both sides by 2.74

$$[CH_3COOH] = \frac{50}{2.74} = 18.25 \text{ mM}$$ Substitute $[CH_3COOH]$ for 18.25 mM

$$[CH_3COO^-] = 50 - 18.25 = 31.75 \text{ mM}$$

Calculation 4.2 Calculation of the relative concentrations of salt and acid for 50 mM ethanoic acid buffer.

These values can then be substituted into Equation 4.7 to calculate the ionic strength, see Calculation 4.3.

$$1 = \frac{1}{2}[0.031\ 75(1)^2 + 0.031\ 75(-1)^2] = 0.3175\ M = 31.75\ mM$$

Calculation 4.3   Expression for the calculation of the ionic strength of 50 mM ethanoic acid buffer.

However, the apparent pK_a (pK_a^*) can change according to ionic strength (and with temperature). The variation of pK_a^* with I, for ionic strengths up to 0.1 M is approximated by (Equation 4.8).

$$pK_a^* = pK_a + 0.509(2z + 1)\left[\frac{\sqrt{I}}{(1 + 1.6\sqrt{I})}\right]$$

Equation 4.8   Expression for the calculation of the pK_a for ionic strengths up to 0.1 M.
For ionic strength between 0.1 and 0.5 M is approximated by Equation 4.9, where 0.509 is the value of the Debye-Hückel constant for water at 25 °C and 1.6 is an empirical constant.

$$pK_a^* = pK_a + 0.509(2z + 1)\left[\frac{\sqrt{I}}{(1 + \sqrt{I})} - 0.2I\right]$$

Equation 4.9   Expression for the calculation of the pK_a for ionic strengths between 0.1 and 0.5 M.

For the ethanoic acid-based buffer system described above the pK_a^* is 4.68 (compared to the true or thermodynamic pK_a of 4.76) at I = 0.033 M.

Thus, the calculation needs to be adjusted for this difference, leading to the $[CH_3COOH] = 16.18$ mM and the $[CH_3COO^-] = 33.82$ mM, with I = 33.82 mM.

Key Concept: Debye–Hückel Theory

The Debye–Hückel theory is a model describing electrostatic interactions in an electrolyte solution which is formed when electrolytes (e.g. salts) are dissolved into a polar solvent or solvent mixture and they dissociate into ions either partially or completely.

4.4   Change in pH with Temperature

As mentioned above, and as all equilibrium constants vary with temperature, the pK_a of a buffer varies with temperature according to the Van't Hoff equation (Equation 4.10).

$$\frac{d\ln(K)}{\Delta T} = \frac{\Delta H}{RT^2}$$

Equation 4.10   Expression for the Van't Hoff equation.
ΔT is the change in temperature, ΔH is the enthalpy, R is the gas constant and T is the temperature in K.

The change in enthalpy is also temperature-dependent, being controlled by the heat capacity (Equation 4.11).

$$\Delta C_p = \frac{\delta\Delta H}{\Delta T}$$

Equation 4.11   Expression for the heat capacity.
$\delta\Delta H$ is the change in enthalpy and ΔT is the change in temperature.

Table 4.1 Temperature dependence of pH for some commonly used buffers [5–9].

Buffer	pK_a (20 °C)	ΔpK_a (°C)
Ethanoic acid	4.76	+0.0002
Succinic acid	5.64	+0.0002
L-Histidine	6.14	−0.017
MES	6.15	−0.011
Bis-tris	6.48	−0.017
ADA	6.60	−0.011
PIPES	6.80	−0.0085
ACES	6.90	−0.020
Imidazole	7.01	−0.021
BES	7.15	−0.016
MOPS	7.20	−0.013
Phosphoric acid	7.21	−0.0022
TES	7.50	−0.020
HEPES	7.55	−0.014
TEA	7.86	−0.020
Tricine	8.15	−0.021
Tris	8.30	−0.031
Bicine	8.35	−0.018
Glycylglycine	8.40	−0.028
TAPS	8.56	−0.024
Diethanolamine	9.01	−0.026
CHES	9.55	−0.023
CAPS	10.64	−0.028

Although over relatively small temperature ranges (such as usual laboratory temperature changes) the enthalpy may be assumed to be constant. This means that the pH of buffers is temperature dependent, meaning that the temperature at which the buffer is to be used should be considered when preparing the buffers at temperatures that differ from this. The temperature dependence of pH for several common buffers is shown in Table 4.1.

4.5 Choice of Buffer

Several factors are important when choosing a buffer system in which to conduct enzyme kinetic studies. Of course, the pK_a of the buffer is critical for maintaining the pH and so a buffer with a pK_a within 1 unit of the desired pH should ideally be selected. The buffer should be water soluble to allow concentrated stock solutions to be made (whilst remembering that the pH may change with dilution, so the correct pH at the relevant concentration should be ensured). The buffer should not interfere with known biological processes and should be stable and resistant to degradation.

Table 4.2 Additional considerations in the choice of buffer. Metal ion interactions are illustrated [20] and radical reactions are illustrated [21–23]. Further details of Good's buffers are illustrated [5, 24, 25].

Buffer	Additional considerations
Ethanoic acid	
Succinic acid	Chelates metal ions
L-Histidine	Histidine solutions have been observed to change color under high temperatures and acidic pH.
MES	Substitute for cacodylate, citrate, and malate buffers [10]. Inhibits connexin channels [11]. Binds human liver FAB protein [12].
Bis-tris	Binds Cu^{2+} and Pb^{2+}. Substitute for cacodylate [10]. Interacts with human liver FAB and affects protein dynamics. Not suitable for BCA assay
ADA	Strong binding to many divalent metal ions. Absorbs UV light between 0.1 and 260 nm. Not suitable for BCA assay.
PIPES	May form radicals, so not suitable for redox reactions [13]. Suitable for BCA assay.
ACES	Binds Cu^{2+} and Mg^{2+}. Significant absorption of UV light at 230 nm. Inhibits GABA receptor binding [14].
Imidazole	Chelates various divalent cations.
BES	Binds Cu^{2+} and Co^{2+} weakly. Suitable for BCA assay. Binds DNA and interferes with restriction enzymes [15].
MOPS	Filter sterilization is required. Suitable for BCA assay.
Phosphoric acid	Phosphates react with calcium producing insoluble calcium phosphate. Phosphate ions can inhibit the activity of some enzymes, for example, carboxypeptidase, carboxylase, and phosphoglucomutase.
TES	Binds strongly to Cr^{3+}, Fe^{3+} and weakly to Co^{2+}, Ni^{2+}, Cu^{2+} and Zn^{2+}. Suitable for BCA assay.
HEPES	May form radicals, so not suitable for redox reactions. Binds DNA and affects restriction enzyme function.
TEA	Acts as a surfactant.
Tricine	Binds Cu^{2+} strongly, Ca^{2+} and Mn^{2+} moderately, and Mg^{2+} weakly. Substitute for barbital [16]. Undergoes photooxidation by flavins and flavoproteins [17].
Tris	Reacts with aldehydes, common metals, various enzymes, and DNA [18]. Not suitable for BCA assay. Not suitable for most cell cultures.
Bicine	Binds strongly to Mg^{2+}, Ca^{2+}, Co^{2+}, Fe^{3+} and Cu^{2+}, weakly to Mn^{2+}
Glycylglycine	Expensive buffer that only works well above pH 8.0, complexes with cations.
TAPS	Binds strongly to Cr^{3+}, Fe^{3+} and Cu^{2+}.
Diethanolamine	Acts as a surfactant.
CHES	High affinity for iodoacetate binding site of liver alcohol dehydrogenase [19].
CAPS	Weak binding to most metals. Suitable for BCA assay.

There should be minimal effects of salts, which may create reactions that may cause complications in enzyme assays. The selected buffer must not interact with the protein used in the assay and it must not inhibit the enzyme or function as a substrate. Clearly, buffers that can act as chelating agents cannot be used with enzymes that are dependent upon metal ions or activity. As well as not interfering with the enzyme, the chosen buffer must not interfere with the assay detection system (e.g. the buffer should not absorb either in the visible or in the UV region). Often, large counter ions are preferred, since they are less likely to interact with proteins. Ligand binding and catalysis

Table 4.3 Composition of three component buffer systems covering a range of pH values (4.5–9.5) with constant ionic strength.

Buffers	pK_a (I = 0.1, 30 °C)	Concentration (M) for I = 0.1
Ethanoic acid	4.64	0.05
MES	6.02	0.05
N-ethylmorpholine	7.68	0.1
Ethanoic acid	4.64	0.05
MES	6.02	0.05
TEA	7.78	0.1
Ethanoic acid	4.64	0.05
MES	6.02	0.05
Tris	8.00	0.1
MES	6.02	0.052
TAPSO	7.49	0.052
Diethanolamine	8.88	0.1
Ethanoic acid	4.64	0.1
Bis-tris	6.32	0.05
TEA	7.76	0.05
MES	6.02	0.1
N-ethylmorpholine	7.68	0.051
Diethanolamine	8.88	0.051
ACES	6.65	0.1
Tris	8.00	0.052
Ethanolamine	9.47	0.052
Succinic acid	5.38	0.033
Imidazole	6.97	0.044
Diethanolamine	8.88	0.044

may often be affected by ionic strength, and it is essential to maintain ionic strength constant across the pH range when conducting pH-dependent studies. This is particularly important as the ionic strength of many buffers changes with pH, see Table 4.2 for additional considerations for typical buffers.

One approach to avoid variation in ionic strength as the buffer ionizes with changing pH is to use mixtures of buffers. Table 4.3 describes three component buffer systems that produce almost constant ionic strength when used across the pH range from 4.5 to 9.5. Since the ionic form of the components changes with pH, it is important to check that none of the ionic species perturbs molecular recognition. This may be accomplished by characterizing the system over a range of pH values and varying the buffer concentrations. The results are then compared with those observed when the ionic strength is varied by addition of an inert electrolyte, such as NaCl.

4.6 Characteristics of Ionizing Groups in Proteins

There are three physical properties that are commonly used to identify ionizing groups. These are pK_a, the enthalpy of ionization (ΔH_{ion}), and the effect of solvent. Typical values for pK_a and enthalpy of ionization are shown in Table 4.4. Note that the properties of ionizing groups in proteins are highly dependent upon the local environment, and so sometimes the pK_a values may be perturbed.

Table 4.4 Properties of ionizing groups in proteins.

Group	Ionization reaction	pK_a	ΔH_{ion} (kJ/mol)
α-carboxyl (C terminus)	$COOH = COO^- + H^+$	2.0–5.5 (Often 3.0–3.2)	−6.3 to 6.3
β or γ-carboxyl (Asp or Glu)		2.0–5.5 (Often 3.0–5.0)	
Imidazolium (His)	$ImH^+ = Im + H^+$	5.0–8.0 (Often 5.5–7.0)	28.9–31.4
α-amino (N terminus)	$NH_3^+ = NH_2 + H^+$	7.5–8.5	41.8 – 54.4
ε-amino (Lys)		9.5–10.6	
Sulfhydryl (Cys)	$SH = S^- + H^+$	8.0–11.0 (Often 8.0–8.5)	27.2–29.3
Phenolic hydroxyl (Phe)	$OH = O^- + H^+$	9.0–12.0 (Often 9.8–10.5)	25.1
Guanidinium (Arg)	$GuH^+ = Gu + H^+$	11.6–12.6	50.2–54.4
Water coordinated to metal (Mg^{2+}, Zn^{2+})	$H_2O \cdots Me^{2+} = OH^- \cdots Me^{2+} + H^+$	8.0–10.0	46.0–54.4

4.7 Effect of pH on Enzyme Activity

To conduct experiments which characterize the effects of pH on enzyme activity, it is essential that the assay system is stable over the pH range to be studied. Distinction must be made between reversible and irreversible changes in activity with pH. The aim is to monitor reversible processes, which are due to ionization, and to avoid irreversible effects that invalidate the results. Stability of assay systems often varies with temperature, ionic strength, the chemical nature of the buffer, the concentration of reagents, the concentration of substrates and cofactors, and the concentration of protein. In order to check the stability of an assay system that is usually used at pH 7, over a pH range between 4 and 9, the stability of the enzyme and substrate should be measured by preincubating them separately at the various pH values and the assay temperature, for a period of time that is at least twice as long as the usual assay time. Any irreversible inactivation can be observed following the return to pH 7 and conducting the assay in the usual way. It is important to use more than one buffer at each pH value to check for effects caused by the chemical nature of the buffer. The buffers used should be evaluated for inhibition by both the protonated and non-protonated forms, by measuring activity at $pH = pK_a \pm 0.5$.

The activity of enzymes may vary by several orders of magnitude over the pH range used. Thus, it may be necessary and important to vary the concentration of enzyme in the assay to obtain a measurable rate. Of course, this approach may be problematic if activity is not linearly correlated with concentration.

4.7.1 Assumptions Required for the Analysis of pH Dependence

To conduct studies on the pH dependence of enzyme activity, the following assumptions are made: ionizing groups are considered to act as perfectly titrating acids or bases; intermediates are assumed to be at protonic equilibrium since proton transfer is faster than any chemical steps; the rate determining step does not change with pH.

These assumptions are usually valid, but the breakdown of the protonic equilibrium of intermediates may sometimes occur if there is slow dissociation of a substrate or product, which prevents the transfer of protons to or from ionizing groups. The conditions may also be invalid when there is rapid turnover of a substrate, for example, in the situation where the rate-determining step is proton transfer. For these reasons, it is often useful to monitor the pH dependence by utilizing several substrates, differing in kinetic properties, or by using reversible inhibitors that act as substrate analogues.

4.7.2   General Rate Equation for pH Dependence

Consider the general scheme below (Reaction scheme 4.14) with respect to pH dependence.

Reaction scheme 4.14   General reaction scheme for demonstrating the effect of protonation and deprotonation of enzyme catalysis.
K_m represents the Michaelis-Menten constant for the singularly protonated enzyme. K_m' represents the Michaelis-Menten constant for the deprotonated enzyme. K_m'' represents the Michaelis-Menten constant for the doubly protonated enzyme. K_{E1}, K_{E2}, K_{ES1}, K_{ES2} represent the acid dissociation constants for the respective enzyme forms. k_{cat} represents the rate constant for product formation and α and β represent the factors by which the value of k_{cat} is modified by either deprotonation or protonation, respectively.

The rate equation for Reaction scheme 4.14 is given by Equation 4.12, below.

$$v = \frac{\frac{k_{cat}[E]_t[S]\alpha\left(K_{ES1}+[H]+\frac{\beta[H]^2}{K_{ES2}}\right)}{K_{E1}K_m}}{1+\frac{[S]K_{ES1}}{K_{E1}K_m}+\frac{[H]\left(1+\frac{[S]}{K_m}\right)}{K_{E1}}+\frac{[H]^2\left(1+\frac{[S]K_{E2}}{K_mK_{ES2}}\right)}{K_{E1}K_{E2}}}$$

Equation 4.12   Expression for the rate equation for general pH dependence.

Note that in both the reaction scheme and rate equation (Equation 4.12) $K_mK_{E2} = K_m''K_{ES2}$ and $K_mK_{E1} = K_m'K_{ES2}$ and that in all reaction schemes and equations in this section, the charges have been removed for clarity.

In the presence of inhibitor, this general scheme (Reaction scheme 4.14) becomes Reaction scheme 4.15.

Reaction scheme 4.15   General reaction scheme for demonstrating the effect of protonation and deprotonation of enzyme catalysis in the presence of inhibitor.
Note that in this scheme, charges have been removed for clarity. K_{I0}, K_I, and K_{I2} represent the dissociation constants for the inhibitor binding to the deprotonated, singularly protonated, and doubly protonated free enzyme forms, respectively. K_{IES0}, K_{IES}, and K_{IES2} represent the dissociation constants for the inhibitor binding to the deprotonated, singularly protonated, and doubly protonated enzyme within the enzyme-substrate complex, respectively.

The rate equation for Reaction scheme 4.15 is derived assuming that $[I] \gg [E]_t$ and that the binding of inhibitor completely prevents catalysis (Equation 4.13).

$$v = \frac{\dfrac{k_{cat}[E]_t[S]\left(\alpha K_{ES1} + [H] + \dfrac{\beta[H]^2}{K_{ES2}}\right)}{K_{E1}K_m}}{1 + \dfrac{[I]}{K_{I0}} + \dfrac{[S]K_{ES1}\left(1 + \dfrac{[I]}{K_{IES0}}\right)}{K_{E1}K_m} + \dfrac{[H]\left(1 + \dfrac{[I]}{K_I} + \dfrac{[S]}{K_m} + \dfrac{[S][I]}{K_m K_{IES}}\right)}{K_{E1}} + \dfrac{[H]^2\left(1 + \dfrac{[I]}{K_{I2}} + \dfrac{[S]K_{E2}}{K_m K_{ES2}} + \dfrac{[S][I]K_{E2}}{K_m K_{ES2} K_{IES2}}\right)}{K_{E1}K_{E2}}}$$

Equation 4.13   Expression for the rate equation for pH dependence in the presence of an inhibitor.

Now consider the situation where only the deprotonated enzyme can turn over substrate to product, the scheme above simplifies to (Reaction scheme 4.16).

Reaction scheme 4.16   General reaction scheme for enzyme catalysis when only the deprotonated form of the enzyme is catalytically active.

In this case, k_{cat} must be slow so that the system conforms to Michaelis-Menten kinetics, and the dissociation constants are related: $K_E K_m' = K_{ES} K_m$, which means that if $K_E = K_{ES}$, then $K_m = K_m'$ and there is no pH dependence for binding.

If $K_E \neq K_{ES}$, the pK_a is perturbed upon binding so that $K_m = K_m' K_E / K_{ES}$ and binding of substrate is dependent upon pH. The rate equation is shown below (Equation 4.14).

$$v_H = \frac{k_{cat}[E]_t[S]}{K_m + [S]\left(1 + \frac{[H]}{K_{ES}}\right) + \frac{K_m[H]}{K_E}}$$

Equation 4.14   Expression for the rate equation for Reaction scheme 4.14 where $K_E \neq K_{ES}$.

This rate equation can be used to predict the pH dependence of k_{cat}, which is observed when $[S] \gg K_m$ and which leads to the relationship shown below (Equation 4.15).

$$(V_{max})_H = [E]_t (k_{cat})_H = \frac{k_{cat}[E]_t K_{ES}}{(K_{ES} + [H])}$$

Equation 4.15   Rate equation for pH dependence of V_{max}.

The denominator demonstrates that the pH dependence of k_{cat} follows the ionization constant of the ES complex, K_{ES}.

An expression for the pH dependence of K_m can be obtained by expressing the rate equation above in terms of the simple Michaelis-Menten form as shown below (Equation 4.16).

$$v_H = \frac{k_{cat}[E]_t[S]}{\frac{K_m K_{ES} + \frac{K_m K_{ES}[H]}{K_E}}{K_{ES} + [H]} + [S]}$$

Equation 4.16   Expression following rearrangement of Equation 4.13 to the same form as the Michaelis-Menten equation.

Where the expression for K_m, becomes Equation 4.17.

$$(K_m)_H = \frac{K_m K_{ES} + \frac{K_m K_{ES}[H]}{K_E}}{K_{ES} + [H]}$$

Equation 4.17   Expression for the pH dependence of K_m.

This also shows that the pH dependence of K_m follows ionization of the ES complex.

The pH dependence of k_{cat}/K_m can be obtained by the ratio of the expressions for k_{cat} and K_m as shown in Equation 4.18.

$$\left(\frac{k_{cat}}{K_m}\right)_H = \frac{\left(\frac{k_{cat}}{K_m}\right) K_E}{K_E + [H]}$$

Equation 4.18   Expression for the pH dependence of k_{cat}/K_m.

In this case, it can be seen from the denominator that the pH dependence follows the ionization of the free enzyme (and free substrate, which has been assumed does not ionize).

These relationships provide a general rule that the pH dependence of a process A → B, follows the ionization of A (Table 4.5).

Table 4.5 pH dependence and assignment of pK_a values.

pH dependence of	Process	pH dependence follows ionization of
k_{cat}	$ES \rightarrow E + P$	ES
K_m	$ES \rightarrow E + S$	ES
$1/K_m$	$E + S \rightarrow ES$	E and S
k_{cat}/K_m	$E + S \rightarrow E + P$	E and S
K_I	$EI \rightarrow E + I$	EI
$1/K_I$	$E + I \rightarrow EI$	E and I
K_{IES}	$IES \rightarrow IE + S$	IES
$1/K_{IES}$	$IE + S \rightarrow IES$	IE and S

It is important to note that the pH dependence of dissociation constants (K_m, K_I, K_{IES}) follows ionization of different molecules to the pH dependence of association constants ($1/K_m$, $1/K_I$, $1/K_{IES}$). The reason for this is that the reactants are different. However, it does mean that some pK_a values can be measured in different ways. This can help in the assignment of pK_a values, for example, ascertaining whether the pH dependence of $1/K_m$ is due to ionization of the free enzyme or free substrate. It can also be used to gain an additional measure of the observed pK_a value, to improve confidence in the calculated value.

Table 4.5 illustrates that the pH dependence of K_m and K_I tends to follow the ionization of groups involved in binding of the substrate and inhibitor. Additional pK_a values observed in the profile for k_{cat} or k_{cat}/K_m can potentially identify groups involved in catalysis.

4.7.3 Fitting pH Dependence

For substrate dependence, ideally, [S] and pH should be varied in a single experiment and measured rates fitted to Equation 4.12, or to simplified models derived from it. For inhibitor dependence, ideally [S], [I] and pH should be varied in a single experiment and measured rates fitted to an expression for the general scheme in the presence of inhibitor (Equation 4.13), but this is often not possible due to technical limitations.

4.7.3.1 Double Ionizing Systems
Consider the reaction scheme for a double ionizing system, Reaction scheme 4.17, below.

$$H_2A \; \underset{}{\overset{K_a}{\rightleftharpoons}} \; HA^- + H^+ \; \underset{}{\overset{K_2}{\rightleftharpoons}} \; A^{2-} + 2H^+$$

Reaction scheme 4.17 General reaction scheme for a double enzyme ionization.

In a doubly ionizing system such as this, a parameter L (such as a rate or a binding constant) may be defined, so that the rate, or binding is a function of L and the concentration (Equation 4.19).

$$L_H[A]_0 = L_{H_2A}[H_2A] + L_{HA}[HA] + L_A[A]$$

Equation 4.19 Expression for the parameter, L_H, of a doubly ionizing system defined as a function of concentration.
For H_2A, $L = L_{H_2A}$; for HA, $L = L_{HA}$; and for A, $L = L_A$. Charges are omitted for clarity.

Where the expressions for [H_2A], [HA], and [A] are shown in Equation 4.20.

$$[H_2A] = \frac{[A]_0}{1 + \frac{K_1}{[H]} + \frac{K_1 K_2}{[H]^2}}$$

$$[HA] = \frac{[A]_0}{1 + \frac{[H]}{K_1} + \frac{K_2}{[H]}}$$

$$[A] = \frac{[A]_0}{1 + \frac{[H]}{K_2} + \frac{[H]^2}{K_1 K_2}}$$

Equation 4.20   Expressions for [H_2A], [HA], and [A] are used in Equation 4.19.
Charges are omitted for clarity.

Following substitution from Equation 4.20, Equation 4.19 then becomes Equation 4.21, below.

$$L_H = \frac{L_{H_2A}[H]^2 + L_{HA}K_1[H] + L_A K_1 K_2}{K_1 K_2 + K_1[H] + [H]^2}$$

Equation 4.21   Expression for the parameter, L_H, of a doubly ionizing system substituting expressions from Equation 4.20 in Equation 4.19.

When $L_{H_2A} = L_A = 0$, then the value of L_H is obtained from the expression shown in Equation 4.22.

$$L_H = \frac{L_{HA}}{1 + \frac{[H]}{K_1} + \frac{K_2}{[H]}}$$

Equation 4.22   Expression for the equation of a bell-shaped pH dependence curve for a double enzyme ionization.

Equation 4.22 defines a bell-shaped curve, often associated with enzyme pH optimum (Figure 4.2). The bell-shaped curve has a maximum at $pH = \frac{(pK_1 + pK_2)}{2}$.

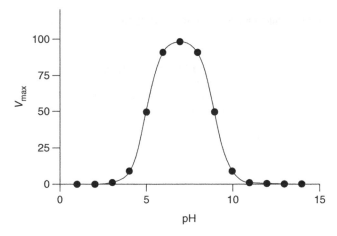

Figure 4.2   Bell-shaped pH-dependence curve.

4.7.3.2 Singularly Ionizing Systems

$$H_2A \xrightleftharpoons{K_a} HA^- + H^+$$

In a simpler, single ionization process such as the one in Reaction scheme 4.18, the rate or affinity can be defined as a function of L as shown in Equation 4.23 and yields a plot of rate or affinity versus pH as shown in Figure 4.3.

$$H_2A \xrightleftharpoons{K_a} HA^- + H^+$$

Reaction scheme 4.18   General reaction scheme for a single enzyme ionization.

$$L_H[A]_0 = L_{HA}[HA] + L_A[A]$$

Equation 4.23   Expression for the parameter of a singly ionizing system defined as a function of concentration.

The expressions for the concentrations of HA and A are given in Equation 4.24, below.

$$[HA] = \frac{[A]_0}{1 + \frac{K_a}{[H]}}$$

$$[A] = \frac{[A]_0}{1 + \frac{[H]}{K_a}}$$

Equation 4.24   Expression for the definitions of [HA] and [A] used in Equation 4.23.
Charges are omitted for clarity.

Following substitution from Equation 4.24 into Equation 4.23 gives Equation 4.25, below.

$$L_H = \frac{L_{HA}[H] + L_A K_a}{K_a + [H]}$$

Equation 4.25   Expression for the parameter for a single ionization by substituting the parameters defined in Equation 4.24 into Equation 4.23.

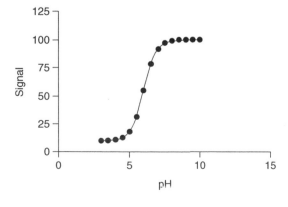

Figure 4.3   pH profile assuming a single ionization event.
This describes how the activity varies with pH for any system where only a single ionization occurs over the pH range studied. The pK_a is given by the point of inflection of the curve.

4.8 Effect of Solvent and Ionic Strength

Note that often solvents are added to enzyme solutions. For example, compounds used in drug discovery are often dissolved in dimethyl sulfoxide (DMSO). DMSO is a polar aprotic solvent that is used to dissolve and store compounds and so is often added to the assays that are used for characterizing enzymes and for hit identification. Increasing solvent polarity stabilizes ions and so may affect the pK_a. DMSO may therefore influence pK_a, for example, the pK_a of water is 14.0 in pure water, but when dissolved in DMSO it is 32.

For a neutral acid, ionization is favored by increasing polarity. The addition of a non-polar solvent will decrease pK_a and the addition of a polar solvent will increase pK_a. Cationic acids are less sensitive to solvent polarity since there is no change in charge of ionization. Some ionizing groups on proteins are neutral acids, with others being cationic. Characterizing the effect of solvent polarity can help in the identification of ionizing groups. However, adding solvent also alters the pK_a of the buffer and so the pH of the solution. Hence, the behavior of protein pK_a values should ideally be measured in neutral acid and cationic buffers and in the presence and absence of solvent (at concentrations that do not denature or inhibit the protein). The pH should be measured before addition of the solvent with rates determined in buffers with and without the solvent.

References

1 Yadav JK, Prakash V. Stabilization of α-amylase, the key enzyme in carbohydrates properties alterations, at low pH. *International Journal of Food Properties*. 2011;14(6):1182–96.
2 Sky-Peck HH, Thuvasethakul P. Human pancreatic alpha-amylase. II. Effects of pH, substrate, and ions on the activity of the enzyme. *Annals of Clinical and Laboratory Science*. 1977;7(4):310–7.
3 Wang R, Edrington TC, Storrs SB, Crowley KS, Ward JM, Lee TC, et al. Analyzing pepsin degradation assay conditions used for allergenicity assessments to ensure that pepsin susceptible and pepsin resistant dietary proteins are distinguishable. *PLoS One*. 2017;12(2):e0171926.
4 Campos LA, Sancho J. The active site of pepsin is formed in the intermediate conformation dominant at mildly acidic pH. *FEBS Letters*. 2003;538(1–3):89–95.
5 Good NE, Winget GD, Winter W, Connolly TN, Izawa S, Singh RMM. Hydrogen ion buffers for biological research. *Biochemistry*. 1966;5(2):467–77.
6 Haynes WM. *CRC Handbook of Chemistry and Physics*. 93: Taylor & Francis; 2012.
7 Goldberg RN, Kishore N, Lennen RM. Thermodynamic quantities for the ionization reactions of buffers. *Journal of Physical and Chemical Reference Data*. 2002;31(2):231–370.
8 Machado CMM, Gameiro P, Soares HMVM. Complexation of M–(buffer)$_x$–(OH)$_y$ systems involving divalent ions (cobalt or nickel) and zwitterionic biological buffers (AMPSO, DIPSO, TAPS and TAPSO) in aqueous solution. *Journal of Solution Chemistry*. 2008;37(5):603–17.
9 Scheller KH, Abel TH, Polanyi PE, Wenk PK, Fischer BE, Sigel H. Metal ion/buffer interactions. Stability of binary and ternary complexes containing 2-[bis(2-hydroxyethyl)amino]-2(hydroxymethyl)-1,3-propanediol (Bistris) and adenosine 5′-triphosphate (ATP). *European Journal of Biochemistry*. 1980;107(2):455–66.
10 Scopes RK. *Protein Purification: Principles and Practice*: Springer New York; 2013.
11 Bevans CG, Harris AL. Regulation of connexin channels by pH. Direct action of the protonated form of taurine and other aminosulfonates. *Journal of Biological Chemistry*. 1999;274(6):3711–9.

12 Long D, Yang D. Buffer interference with protein dynamics: a case study on human liver fatty acid binding protein. *Biophysical Journal.* 2009;96(4):1482–8.

13 Baker CJ, Mock NM, Roberts DP, Deahl KL, Hapeman CJ, Schmidt WF, et al. Interference by Mes [2-(4-morpholino)ethanesulfonic acid] and related buffers with phenolic oxidation by peroxidase. *Free Radical Biology and Medicine.* 2007;43(9):1322–7.

14 Tunnicliff G, Smith JA. Competitive inhibition of γ-aminobutyric acid receptor binding by *N*-2-hydroxyethylpiperazine-*N'*-2-e-ethanesulfonic acid and related buffers. *Journal of Neurochemistry.* 1981;36(3):1122–6.

15 Wenner JR, Bloomfield VA. Buffer effects on EcoRV kinetics as measured by fluorescent staining and digital imaging of plasmid cleavage. *Analytical Biochemistry.* 1999;268(2):201–12.

16 Monthony JF, Wallace EG, Allen DM. A non-barbital buffer for immunoelectrophoresis and zone electrophoresis in agarose gels. *Clinical Chemistry.* 1978;24(10):1825–7.

17 Grande HJ, van der Ploeg KR. Tricine radicals as formed in the presence of peroxide producing enzymes. *FEBS Letters.* 1978;95(2):352–6.

18 Brignac PJ, Mo C. Formation constants and metal-to-ligand ratios for tris(hydroxymethyl) aminomethane-metal complexes. *Analytical Chemistry.* 1975;47(8):1465–6.

19 Syvertsen C, Mckinley-Mckee JS. Affinity labelling of liver alcohol dehydrogenase. Effects of pH and buffers on affinity labelling with iodoacetic acid and (*R,S*)-2-bromo-3-(5-imidazolyl)propionic acid. *European Journal of Biochemistry.* 1981;117(1):165–70.

20 Ferreira CMH, Pinto ISS, Soares EV, Soares HMVM. (Un)suitability of the use of pH buffers in biological, biochemical and environmental studies and their interaction with metal ions – a review. *RSC Advances.* 2015;5(39):30989–1003.

21 Hicks M, Gebicki JM. Rate constants for reaction of hydroxyl radicals with Tris, Tricine and Hepes buffers. *FEBS Letters.* 1986;199(1):92–4.

22 Shiraishi H, Kataoka M, Morita Y, Umemoto J. Interactions of hydroxyl radicals with tris (hydroxymethyl) aminomethane and good's buffers containing hydroxymethyl or hydroxyethyl residues produce formaldehyde. *Free Radical Research Communications.* 1993;19(5):315–21.

23 Kirsch M, Lomonosova EE, Korth H-G, Sustmann R, de Groot H. Hydrogen peroxide formation by reaction of peroxynitrite with HEPES and related tertiary amines: IMPLICATIONS FOR A GENERAL MECHANISM. *Journal of Biological Chemistry.* 1998;273(21):12716–24.

24 Ferguson WJ, Braunschweiger KI, Braunschweiger WR, Smith JR, McCormick JJ, Wasmann CC, et al. Hydrogen ion buffers for biological research. *Analytical Biochemistry.* 1980;104(2):300–10.

25 25. Good NE, Izawa S. 3 Hydrogen Ion Buffers. *Methods in Enzymology.* 24: Academic Press; 1972. p. 53–68.

5

Steady-state Assays and their Design

5.1 Introduction

The aim of many assays, especially in drug discovery, is to understand the effect of compounds on the rate of substrate consumption or product formation. This is most easily determined when the enzyme-bound intermediates are not changing – a period known as the steady-state. This is usually accomplished by setting substrate concentrations well above the enzyme concentration and measuring the initial rate of the enzyme-catalyzed reaction. The following chapter will describe both the pre-steady-state period and describe the development of steady-state assays.

5.2 The Pre-steady State

Before the steady-state can be reached, following mixture of an enzyme with its substrate and or inhibitor, the concentration of enzyme-bound intermediates must increase from zero to their steady-state levels. The period during which the concentration of the enzyme-bound intermediates is increasing is called the pre-steady state. The pre-steady state is extremely useful for characterizing the chemical mechanism of catalysis since this gives direct information on the intermediates as they accumulate (as the rate of breakdown is less than rate of formation). Some of the key differences between pre-steady-state and steady-state assays are shown in Table 5.1.

The experimental approaches used for the study of pre-steady state reactions depend upon the timescale of the reaction under study (Table 5.2).

Laboratory Guide to Enzymology, First Edition. Geoffrey A. Holdgate, Antonia Turberville, and Alice Lanne.
© 2024 John Wiley & Sons, Inc. Published 2024 by John Wiley & Sons, Inc.

Table 5.1 Key differences between steady-state and pre-steady state.

Steady-state	Pre-steady state
Assumes $[S] \gg [E]$	Uses large quantities of enzyme, often stoichiometric with substrate
Ignores intermediates	Detects intermediates in the reaction, their chemical nature, their number, and order
No special equipment required; analysis is fairly easy	Often requires special equipment to measure rapid reactions
Measures equilibrium constants	Measures rate constants
Information on kinetic mechanism – order of addition of substrates and release of the products	Limited ability to resolve multiple steps in complex reactions
	Not limited to enzymes (protein-ligand binding, DNA folding, etc.)

Table 5.2 Time scale for each pre-steady state technique.

Time scale	Device
s	Hand mixing
ms	Stopped flow, chemical quench flow, freeze quench
μs	Temperature jump
ns	Fluorescence depolarization, magnetic resonance
ps	Flash photolysis

An example of the use of pre-steady state kinetics is in the measurement of enzyme concentration via active site titration. The functional enzyme concentration is often different from the protein concentration of the enzyme solution. This may arise because of impurities in the enzyme solution, or in preparations that are almost 100% pure, due to failure of all the enzymes to function correctly (e.g. because of the presence of unfolded protein). Hence, it is important to be able to measure the functional enzyme concentration that is used in the assay. Two methods often are employed to measure the functional enzyme concentration – active site titration, see Figure 5.1 and estimation from tight-binding inhibition (Section 6.6).

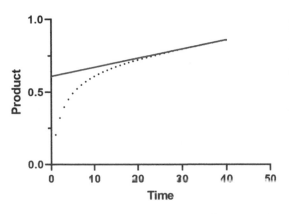

Figure 5.1 Pre-steady state burst kinetics. This figure describes the behavior of an enzyme catalyzed reaction initiated by mixing enzyme with substrate. The amount of the enzyme-product intermediate rapidly increases, in a burst, until the reaction reaches a steady-state phase, indicated by the red, solid line.

Active site titration involves measuring the initial burst of product formation, which occurs when the enzyme is initially mixed with substrate. This occurs when an enzyme-bound intermediate accumulates during the reaction. The first mole of substrate reacts with the enzyme to form stoichiometric amounts of enzyme-bound intermediate and product, but the subsequent reaction is limited by the breakdown of the intermediate to release free enzyme, Reaction scheme 5.1.

$$E + S \underset{k_{-1}}{\overset{k_1}{\rightleftharpoons}} ES \xrightarrow{k_2} ES' + P_1 \xrightarrow{k_3} E + P_2$$

Reaction scheme 5.1 **Reaction scheme for a reaction which will lead to a pre-steady state burst as product P$_1$ is released, before the subsequent rate-limiting reaction step, controlled by k_3, releases P$_2$ and generates free enzyme.**

In this reaction scheme, it is assumed that $k_2 \gg k_3$ to produce the initial burst of P$_1$. The magnitude of the burst is given by Equation 5.1.

$$\text{Burst} = [E]_t \left\{ \frac{\left(\frac{k_{cat}}{k_3} \right)}{\left(1 + \frac{K_m}{[S]_0} \right)} \right\}^2$$

where

$$k_{cat} = \frac{k_2 k_3}{(k_2 + k_3)}$$

and

$$K_m = \frac{k_3 (k_{-1} + k_2)}{k_1 (k_2 + k_3)}$$

Equation 5.1 **Expression for the pre-steady state burst equation, relating to Reaction scheme 5.1.** The parameter values k_{cat} and K_m are the catalytic constant and the Michaelis constant, respectively, see Section 5.6.2.

5.3 Steady-state Assays

Most enzyme assays are carried out in the steady-state. The steady-state usually refers to a situation in which a quantity is constant, that is in a steady-state, because the rate of its formation is equal to the rate of its breakdown. The steady-state in enzyme assays refers to the concentration of enzyme-bound intermediates. There can be multiple enzyme-bound intermediates. The most common intermediate is the enzyme-substrate complex (ES). At steady-state, the concentration of these enzyme-bound intermediates is constant. Prior to the steady-state, the concentration of enzyme-bound intermediates increases, as described in Section 5.2. The steady-state is used more frequently to study enzymes, as these measure the catalytic activity of enzymes under conditions more closely related to those in the cell.

The steady-state of enzyme assays is an approximation since the substrate is depleted during the reaction. However, if the measurements of rate are confined to a period during which the substrate concentration is approximately constant, then the approximation is valid. The steady-state does, however, give only indirect information on intermediates, because it is the decrease in substrate or increase in product concentration that is measured, and as mentioned above, intermediates will only be detected if they accumulate.

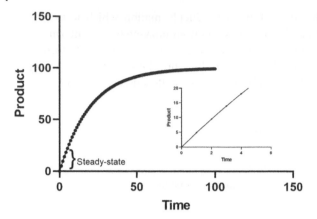

Figure 5.2   **Steady-state progress curve.**

A plot of the concentration of product accumulated in an enzyme-catalyzed reaction versus the time of the reaction is known as the progress curve or time-course of the reaction. This progress curve is initially linear, since the concentration of the enzyme-substrate complex is essentially unchanged, and thus the rate of product formation remains constant. However, after longer periods of time, the progress curve becomes non-linear, as the rate of product formation falls (Figure 5.2).

Enzyme inhibition assays are often carried out using initial rate conditions where the reaction is approximately at steady-state. If the reaction is not at steady-state, the rate of the reaction may change over time, and it can be difficult to accurately determine the true inhibitory effect of a compound. To ensure that the steady-state condition is met, certain criteria must be fulfilled leading to certain assumptions and simplification of the mathematics behind the kinetics.

- During the initial part of the progress curve, the concentration of intermediates does not change and the [ES] is constant;
- The enzyme concentration is low compared to the concentration of substrate;
- During the initial phase, very little product is formed, so that $[P] \approx 0$.

It is often assumed that measuring reaction rates over a time period in which only 10% or less of the substrate is utilized will lead to valid initial rate estimation. However, care should be employed to consider possible causes of non-linearity to ensure that this approach is reliable, and that the linear portion of the time-course is used for subsequent analysis (Figure 5.2).

The progress curve (or time-course) can be obtained by mixing the enzyme and substrate(s) and measuring product generation over time. The initial velocity (tangent to the progress curve at $t = 0$), where the rate of product accumulation is constant with time (as the enzyme-substrate concentration is in the "steady-state") of the reaction can then be identified and subsequent experiments should be conducted in this linear range, where usually less than around 10% of the substrate has been converted to product. The enzyme concentration and/or the total time of reaction can be adjusted to ensure linearity during the course of the experiments. Steps to adjust the enzyme concentration and identify the linear portion of the time-course can usually be done within the same experiment.

5.4   Assay Development

The goal of most enzyme assay development is to build assays that can measure the response of the enzyme to various concentrations of test compounds. For the assay to do this, considerations must be given to the choice of reagents, the concentrations they are used at, and the design of the experimental protocols to be followed. For assays on any enzyme target, it is essential to ensure that the appropriate enzyme, substrates, any necessary co-factors, and ideally, control inhibitors are available before beginning assay development. This will facilitate design and experimentation and will expedite identification of optimal conditions.

5.4.1 Requirements for Method Development

The following requirements should be considered during method development: identity of the enzyme target, including amino acid sequence, purity, and the amount available for development, assay validation, and subsequent support of screening or mechanistic studies to support structure–activity relationships. Contaminating enzyme activities should have been identified and eliminated if possible.

It is important to identify, source, and acquire natural or synthetic substrates with the appropriate sequence, chemical purity, and adequate availability. Similarly, buffer components, co-factors, and other necessary additives for enzyme activity measurements (such as detergents, inorganic salts, and protease inhibitors) according to literature protocols and/or previous experience with the enzyme or assay system, should be obtained (see Chapter 4 for more detail).

The stability of enzyme activity under the chosen storage conditions and during experiments (e.g. Selwyn's test, see Chapter 3 for more detail) should be measured. It is important to establish batch-to-batch variation for longer-term assays. As well as the enzyme, other important reagents should be checked for stability, including substrates and inhibitors. This may be especially relevant for protein substrates, where methods like those used for verifying enzyme stability can be applied to check the integrity of the substrate.

It is useful to identify and acquire purified alternative enzyme forms (e.g. phosphorylated/non-phosphorylated enzyme or with other post-translational modifications, or other enzyme forms based on different regulatory mechanisms) or clinically relevant mutant enzymes (if available) for comparison with wild-type enzyme, to attempt to identify physiologically relevant forms for inhibition.

It is important to investigate different plate types for their suitability and compatibility with the assay biochemistry and the physical basis for product detection.

5.4.2 Different Types of Enzymes Assays

Several different types of enzyme assays are available for the measurement of steady-state rates (see below).

5.4.2.1 Direct Continuous Assay

Some enzyme-catalyzed reactions result in a change in the properties of the reactants compared to the products, such that these differences may be measured directly (without further treatment) and continuously (with time). Any difference capable of being measured directly may provide the basis for such assays, for example, changes in absorbance, fluorescence, enthalpy, pH, conductivity, viscosity, and optical rotation have all been used. Direct continuous assays are the preferred assay type as they allow the progress curve of the reaction to be followed, which facilitates the measurement of initial rates, and observation of unexpected behavior. Direct continuous assays are particularly valuable when observing activity for non-rapid equilibrium inhibitors (i.e. the progress curve is non-linear in the presence of an inhibitor but is linear in the absence of an inhibitor). This situation, for so-called slow-binding inhibition, is discussed in Section 6.7 [1–3].

5.4.2.2 Indirect Assays

For many enzyme assays, there is not a convenient direct continuous assay that can be used to follow the progress curve. In this situation, indirect assays must be used. These involve treatment of the reaction mixture, such that a measurable product is formed, an increase in sensitivity is produced or a more convenient measure of the rate of reaction is achieved. It may be possible to develop indirect assays that are capable of continuous measurement, but often indirect assays are used as discontinuous or end-point assays.

5.4.2.3 Discontinuous Indirect Assays

These assays involve stopping the reaction after various times and treating the reaction mixture to produce the desired change in properties. This allows the change in [substrate] or [product] to be quantified and the rate to be estimated. It is essential that the method used to stop the reaction does so instantaneously. It is also important to verify that the rate is linear over the time of the assay, which involves stopping the reaction at various times and checking that the time course is still linear at the chosen time point (Figure 5.3) [4–6].

5.4.2.4 Continuous Indirect Assays

Continuous indirect assays involve the treatment necessary for detecting changes in substrate or product concentration being able to monitor the change continuously during the assay. For this type of assay to be valid, the reaction allowing detection should occur rapidly enough so that the enzyme-catalyzed reaction is rate-limiting. It is also essential that the reagents used to monitor the change do not interfere with the activity of the enzyme.

5.4.2.5 Coupled Assays

These types of assays involve one or more additional enzymes that catalyze a reaction of one of the products to generate a compound that can be quantified. Many possible coupled enzyme systems are available. However, there is always a lag period associated with coupled enzyme assays as the concentration of the substrate for the coupling enzyme(s) builds up. This lag period depends on the K_m value(s) for the substrate(s) and on the amount of coupling enzyme(s) (Reaction scheme 5.2). Generally, these problems are overcome by having very large excesses of coupling enzymes. It is necessary to minimize the lag period so that the coupled assay gives an accurate estimate of the rate for the primary enzyme-catalyzed reaction. It is also important to check for inhibition of the coupling enzymes when studying inhibitors directed against the primary enzyme [7, 8].

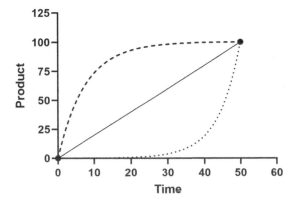

Figure 5.3 **Plot showing the importance of more than one-time point when using discontinuous assays.**
If only one-time point is used, it may be assumed that the rate is linear between time zero and the time at which the product concentration is measured. The dashed and dotted lines show that other behaviors are possible, which can be ruled out if further product measurements are made at additional time points.

$$S \xrightarrow{\quad E \quad} P \xrightarrow{\quad E_{aux1} \quad} Q \xrightarrow{\quad E_{aux2} \quad} R$$

Reaction scheme 5.2 **Scheme for coupled assay with two coupling enzymes, R is measured product.**
Where E is the target enzyme, E_{aux1} and E_{aux2} are the coupling enzymes. S is the substrate for E which gets converted to product, P, which is the substrate for E_{aux1}, this repeats for the second coupling enzyme (E_{aux2}) forming R. R is the product that can be quantified.

5.5 Blank Rates

Enzyme assays commonly suffer from the presence of significant apparent rates of reaction in the absence of one of the components of the assay. It is important to control the presence of these blank rates before undertaking kinetic studies. There are many reasons for the observation of blank rates, including precipitation, contamination, adsorption to assay vessels, non-enzymatic reactions, and others, but it is essential to understand the cause, so that the determined rates used in subsequent analysis are solely due to the enzyme-catalyzed reaction. It is therefore a vital component of assay development to investigate the possible causes of blank rates and to make every effort to minimize and correct them before proceeding to collect experimental data. This often involves measuring the apparent rate in the absence of the enzyme and each substrate individually. Only when a workable signal : noise is obtained should an assay be used for parameter estimation.

5.6 The Assay Development Process

5.6.1 Initial Assay Scoping

The first step in developing an enzyme assay is to assess the linearity of the detection system. This means checking that the signal observed is directly proportional to the product concentration. Sometimes, the products of enzyme reactions can be purchased from commercial suppliers, or they can be generated in the reaction, and characterized. If the product is available, then a standard curve may be constructed to observe the relationship between the product concentration and the measured signal. Plotting the signal obtained versus the concentration of product describes a curve that can then be used to identify the linear portion of detection for the instrument or system used. Subsequent studies must ensure that the product generated, and hence the observed signal falls within this range. This process can allow the identification of artifacts, both generic (to the particular detection system) or specific to the particular product measured. It also allows the sensitivity and the detection limits of the assay to be ascertained.

There are several different terms that may be used to describe the lowest concentration of sample that may be detected by an analytical method used, these are Limit of Blank (LoB), Limit of Detection (LoD), and Limit of Quantitation (LoQ).

LoB represents the highest "apparent" analyte concentration expected to be found when replicates of a blank sample containing no analyte are measured (Equation 5.2).

$$LoB = mean_{blank} + 1.645(SD_{blank})$$

Equation 5.2 **Expression for the limit of blank.**
Where $mean_{blank}$ is the mean of the blank measurements and SD_{blank} is the standard deviation of the blank measurements.

The LoD, which represents the lowest analyte concentration that can be reliably distinguished from the LoB, which is most commonly used, is then calculated (Equation 5.3)

$$LoD = LoB + 1.645(SD_{LCS})$$

Equation 5.3   Expression for the limit of detection.
Where SD_{LCS} represents the standard deviation of a sample containing a low concentration of the analyte.

The LoQ represents the lowest analyte concentration that can be quantitatively detected with a stated accuracy and precision. The LoQ is always greater than or equal to LoD.

Ideally, the relationship between product concentration and signal will be linear over the range of potential concentrations generated during the enzyme reaction. If the relationship is non-linear, it is important to understand the relationship, so that the product concentration generated can be calculated correctly, or that the assay can be interpreted in terms of the ability to identify inhibitors/activators.

If the product response curve is not linear, it is worth spending some time attempting to understand the rationale for non-linear portions. For example, this may require investigations into the substrate concentration used, the pH the assay is run at, the ionic strength, or other conditions that may affect the generation or detection of product.

Once the behavior of the detection system is understood, then initial scoping may be undertaken. This seeks to determine the concentration of enzyme to be used, the accompanying concentration of substrate(s), and the reaction time. Often, these initial conditions may be derived from examining the literature or from previous experience of the system or related systems. At this stage, it may be prudent to undertake some limited laboratory work to confirm the desired activity under the initial conditions identified.

For these initial scoping experiments, it is important that the amount of product generated by the enzyme is within the range of the detection system (it should be remembered that non-linear detection curves will produce non-linear progress curves, if the curvature has not been accounted for). If not, then conditions may need to be adjusted so that the detection sensitivity can be improved. If this cannot be done, an alternative detection system may be required. Non-linear progress curves obtained when the detection system is appropriate and linear, may be a symptom of enzyme instability in the assay. This can be checked by varying the preincubation time and maintaining a constant reaction time. Additionally, approaches such as Selwyn's test (Chapter 3) may be applied. If instability is inferred, then it may be possible to identify conditions that can stabilize the enzyme (alternative buffer systems, plate types or identity and concentration of additives, such as detergents or reducing agents may help). Alternatively, it may be necessary to use an alternative enzyme construct (e.g. a domain of the enzyme instead of the full enzyme) or isozyme.

If curvature in the progress curve is not due to detection system issues or instability of the enzyme, it may arise due to substrate depletion and so a decrease in the ability to form encounter complexes that lead to turnover. It may also occur if the reaction approaches equilibrium, where the reverse reaction competes with the forward reaction. It is possible to check for these effects by decreasing the enzyme concentration and/or by shortening the reaction time. This will allow the development of the [Enzyme] versus reaction time relationship, to balance these conditions to improve the linearity of the progress curves.

The next step is to investigate the substrate dependence of the enzyme. It should be noted that substrate turnover is highly dependent upon the kinetic path and on the concentration of other substrates.

5.6.2 Substrate Dependence

Substrate dependence studies aim to characterize how the enzyme-catalyzed rate of reaction changes in response to the added substrate concentration. This is often the first step in understanding catalysis and for subsequently configuring assays to identify compounds that may modulate enzyme function to be used as therapeutic agents. Whilst the rate of reaction increases with enzyme concentration, it is found experimentally that the rate of reaction with respect to substrate follows a hyperbolic profile, where the rate initially increases and then tends to plateau, Figure 5.4. This so-called saturation kinetic behavior was first characterized by Henri [9] and Michaelis and Menten [10, 11] in the early 1900s and led to the development of the Michaelis–Menten equation, which is now a central relationship in enzyme kinetics.

The Michaelis–Menten equation describes a rectangular hyperbola (Figure 5.4), with an asymptote at very high substrate concentrations, where the rate of reaction approaches a maximum value, termed V_{max}.

The substrate concentration at which the rate is equal to half of the maximum rate represents the Michaelis constant, K_m (Equation 5.4). A protocol on how to determine K_m and V_{max} experimentally can be found in Protocol 5.1 and the process for catalysis is described in Reaction scheme 5.3.

$$v = \frac{V_{max}[S]}{K_m + [S]} = \frac{k_{cat}[E]_t[S]}{K_m + [S]}$$

Equation 5.4 **Expression for the Michaelis–Menten equation.**
The maximum rate, V_{max} is equal to $k_{cat}[E]_t$, where k_{cat} is the catalytic constant and $[E]_t$ is the total enzyme concentration. V_{max} represents the limit of the rate when all the enzyme active sites are saturated with substrate. The K_m is the substrate concentration giving half the maximum rate.

The derivation of the equation is provided assuming rapid equilibrium and steady-state assumptions in Appendix F. It can be seen from inspection of these derivations that the Michaelis constant may be comprised of different combinations of rate constants, depending upon the kinetic mechanism and so it does not always equal the dissociation constant for the ES complex.

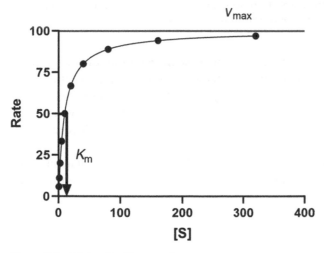

Figure 5.4 **Michaelis–Menten plot.**
The asymptote on the *y*-axis represents the maximum rate, V_{max}, and the K_m value is the concentration of substrate that yields half of the maximum rate. These two parameters are the central parameters of enzyme kinetics.

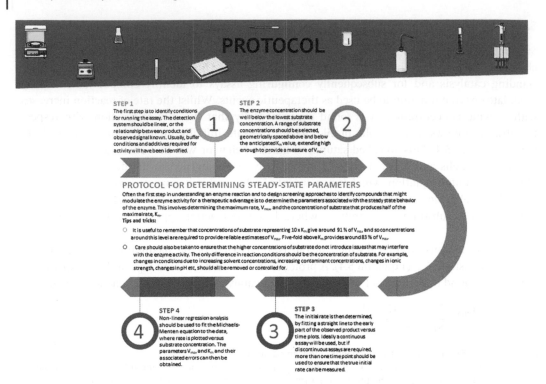

Protocol 5.1   Protocol for determining steady-state parameters for a single substrate enzyme reaction.

$$E + S \xrightleftharpoons{K_s} ES \xrightarrow{k_{cat}} E + P$$

Reaction scheme 5.3   Scheme for a single substrate reaction proposed by Michaelis and Menten.
Where, under rapid equilibrium conditions, the Michaelis constant, K_m, is equal to the dissociation constant for the ES complex, K_s, which describes the behavior shown above (Figure 5.4).

However, it can be considered as an apparent dissociation constant, reflecting the dissociation constant of all the enzyme-bound species. Operationally, the K_m is always the concentration of substrate that yields half of the maximum rate. The catalytic constant, k_{cat}, also known as the turnover number, is a first-order rate constant that reflects the reaction of the enzyme-bound species. It represents the maximum number of substrate molecules converted to product per active site per unit time.

Another important parameter that can be obtained from the Michaelis–Menten equation is the specificity constant, k_{cat}/K_m. When the substrate concentration is much smaller than K_m, the Michaelis–Menten equation reduces to Equation 5.5.

$$v = \frac{k_{cat}[E]_t[S]}{K_m}$$

Equation 5.5   Expression for the form of the Michaelis–Menten equation at low substrate concentrations.

Under these conditions, little enzyme is bound by substrate and so the concentration of free enzyme approximates the total enzyme concentration so that Equation 5.5 becomes Equation 5.6.

$$v = \frac{k_{cat}[E][S]}{K_m}$$

Equation 5.6 **Equation relating the rate of reaction to the free enzyme and substrate concentrations.**

It can be shown that this relationship holds for all concentrations of substrate (see Appendix F). Hence the specificity constant reflects the reactions of the free enzyme and substrate and the value of the apparent second-order rate constant, k_{cat}/K_m determines the specificity for different substrates, hence the name. Another important facet of this parameter is that it cannot be faster than any second-order rate constant in the forward reaction pathway and so it can be used as a lower limit for the association rate constant for enzyme and substrate.

To determine the parameter values associated with the substrate dependence of the enzyme, the substrate is varied over a wide range of values. This allows the V_{max} and K_m to be estimated. It should be noted that K_m' is used for multiple substrate systems, as the value may vary according to the identity and concentration of other substrates (see Section 5.6.2.2). The "true" K_m for a substrate may be estimated by setting the other substrates at concentrations representing full saturation, for example, 100-fold above their K_m values (assuming no complicated kinetics, such as substrate inhibition, described in Chapter 6). As all kinetic constants are dependent upon the conditions, it is important to check these values again if changes are made to the assay during optimization.

Once initial rate conditions have been determined, the substrate concentration should be varied to allow initial rates to be measured for each substrate concentration (Figure 5.5) and the corresponding rates plotted versus concentration to generate a substrate dependence plot (Figure 5.4) which can be used to determine K_m and V_{max} values (see Protocol 5.1). For these values to be correct it is important that initial rate conditions are obeyed. The Michaelis–Menten equation demonstrates that $K_m = [S]$ where $v = \frac{V_{max}}{2}$. Knowledge of the K_m value is important, as this reflects the affinity of the substrate for the enzyme. As mentioned above, K_m may be considered as an apparent dissociation constant, where Equation 5.7 applies.

$$K_m = \frac{[E][S]}{\sum[ES]}$$

Equation 5.7 **Expression for K_m as an apparent dissociation constant.**
The denominator represents the sum of all the bound enzyme species. This shows that K_m is always the concentration of substrate giving half-maximal rate ($V_{max}/2$), as at this rate $[E] = \sum[ES]$.

Figure 5.5 **Time-course for an enzyme reaction at three different substrate concentrations.**
The plot shows the time-course for an enzyme reaction for three different substrate concentrations, where there is curvature. Inset is the measure of the true initial rate (from the tangent to the curves at $t = 0$). These values should be used for determining the kinetic parameters, as they represent steady-state conditions.

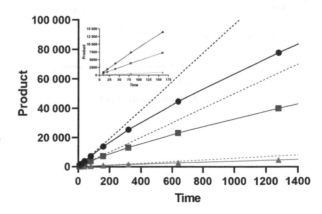

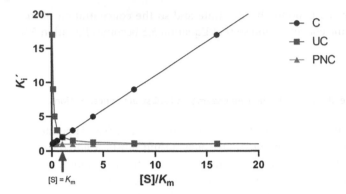

Figure 5.6 **Plot of the change in apparent K_i versus substrate concentration relative to K_m.**
Effect of substrate concentration on the apparent K_i for different mechanisms of inhibition (see Chapter 8).
When the substrate concentration is used at K_m, then the apparent K_i is not more than 2-fold from the true
K_i value. This provides a balanced approach for identifying inhibitors with different mechanisms.

Knowledge of the magnitude of K_m is used in mechanism studies (see Chapter 8) where the
substrate concentration is varied above and below the K_m and the resultant IC_{50} values generated
from concentration-response studies at each of these substrate concentrations provide information
on the type of inhibition mechanism followed. For mono-substrate reactions (and frequently even
for reactions involving multiple substrates), substrate concentrations are often used at their K_m
values for inhibitor screening as this provides a balanced approach to allow identification of all the
different potential mechanisms of inhibition (Figure 5.6) [12].

The effect of the degree of substrate conversion on the IC_{50} values for a competitive, an uncompetitive and a pure noncompetitive inhibitor are shown below (Figure 5.7). Large increases (up
to around 5-fold) in IC_{50} may be observed in the extreme cases of high substrate utilization, compared to those at very low levels, with the effects being most pronounced for compounds showing
uncompetitive inhibition. Thus, it is important to ensure that the substrate conversion remains low
during studies to elucidate kinetic behavior.

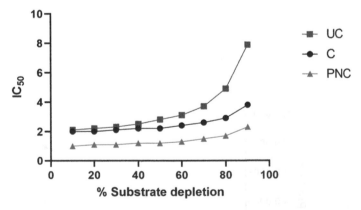

Figure 5.7 **Effect of degree of substrate conversion on measured IC_{50}.**
Increased substrate conversion will increase the resulting IC_{50} value for a given inhibitor. The data were
modelled assuming $K_i = 1.0$ for a competitive inhibitor (C) and an uncompetitive inhibitor (UC) and a pure
non-competitive inhibitor (NC), with no product inhibition, at a substrate concentration equal to K_m. The
rates were calculated assuming the above substrate conversion values for the individual reactions in the
absence of inhibitor. It was assumed that the same time period would then be used to analyze the reactions
in the presence of inhibitor.

5.6.2.1 Non-Michaelian Kinetics

Although many single substrate reactions obey the Michaelis–Menten equation, there are circumstances where it does not adequately describe the variation of rate with [S].

For example, sometimes high concentrations of one or more of the substrates can lead to significant inhibition of the studied enzyme. This is seen as a decrease in the observed rate when the substrate concentration is raised and leads to plots of the shape shown below, Figure 5.8.

Substrate inhibition may occur when a second molecule of substrate can bind to the ES complex to form an inactive ternary complex, ES_2, or sometimes denoted SES, see Figure 5.9.

The rate equation describing this type of behavior is given by Equation 5.8.

$$v = \frac{V_{max}[S]}{K_m + [S] + \frac{[S]^2}{K_{si}}} = \frac{V_{max}}{\left[1 + \left(\frac{K_m}{[S]}\right) + \left(\frac{[S]}{K_{si}}\right)\right]}$$

Equation 5.8   Expression for substrate inhibition.
Where K_{si} represents the dissociation constant for the inhibitory ternary complex (see Appendix F for the derivation).

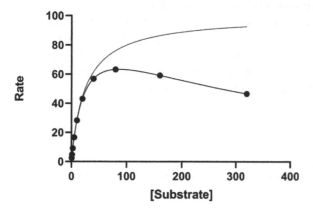

Figure 5.8   Effect of substrate inhibition.
At low substrate concentrations, the kinetics follows simple Michaelis–Menten kinetics (red line), but at high substrate concentrations, there is a significant deviation (black line).

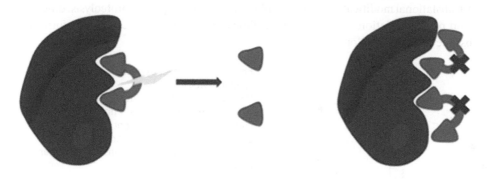

Figure 5.9   Potential mechanism for the cause of substrate inhibition.
At low substrate concentrations, demonstrated by the left-hand side, the enzyme can turn over substrate and follows the usual Michaelian behavior. However, at high substrate concentrations, demonstrated by the right-hand side, the 2-binding sites are filled by different substrate molecules, which are not aligned for turnover and so the reaction is inhibited.

A cautionary note is to check that the appropriate controls have been carried out to rule out artifacts arising from changes in the assay system at high substrate concentrations (substrate solubility, pH, ionic strength, dielectric constant, chelation of metal ions, etc.).

Another common deviation from Michaelis–Menten kinetics leads to sigmoidal substrate dependence curves, which may indicate that the enzyme follows cooperative kinetics. Cooperative kinetics indicates that the binding of one ligand facilitates (positive cooperativity) or impedes (negative cooperativity) the binding of subsequent molecules of the same ligand (see Equation 2.26). The rate equation describing this type of behavior is given in Equation 5.9.

$$v = \frac{V_{max}[S]^h}{K^h + [S]^h}$$

Equation 5.9 Expression for cooperative kinetics.
Where h is the value of the Hill coefficient, with values greater than 1 indicating positive cooperativity and values less than 1 indicating negative cooperativity. K will be the substrate concentration giving $V_{max}/2$ (compare with K_m). Note the similarity to Equation 2.26.

The types of sigmoidal curve obtained are shown below, in Figure 5.10.

Another reason for failure to observe Michaelis–Menten kinetics with single substrate enzymes may be that more than one enzyme present can catalyze the same reaction. If the same reaction is catalyzed by two enzymes (for instance, similar isoforms that have been co-purified, or a single enzyme but with a proportion having altered kinetic properties) in the assay system with different K_m values, then the rate of the overall reaction is determined by Equation 5.10.

$$v = \frac{V_{max}{}^a[S]}{K_m{}^a + [S]} + \frac{V_{max}{}^b[S]}{K_m{}^b + [S]}$$

Equation 5.10 Expression for biphasic kinetics.
Where K_m and V_{max} are the corresponding values for each of the two enzymes, a and b. Note the similarity to Equation 2.25.

The substrate dependence plot (Figure 5.11) appears similar to the plot for negative cooperativity. Where this type of behavior occurs due to different forms of the same enzyme catalyzing the reaction with different K_m or V_{max} values, this may arise due to a number of different reasons, including different post-translational modification (e.g. phosphorylation states), partial proteolysis, different redox status and partial oxidation. Often, it may be difficult to resolve between the two forms by following substrate dependence alone.

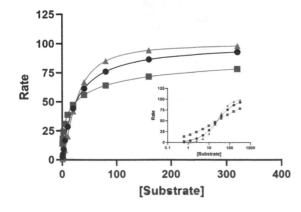

Figure 5.10 Effect of positive and negative cooperativity.
Positive cooperativity (Hill coefficient, $h \gg 1$ (1.5)) is shown by the green line and negative cooperativity (Hill coefficient, $h < 1$ (0.5)) by the pink line. The black line indicates Michaelis–Menten kinetics. Inset are the same curves on a log scale.

Figure 5.11   Effect of biphasic kinetics.
Above is shown a typical substrate dependence plot for an assay containing two enzymes catalyzing the same reaction with different K_m and V_{max} values. The green line appears to be similar to the plot for negative cooperativity (see above) but is actually composed of the sum of the rates for both enzyme reactions (pink and black lines), where the lower, pink line represents an enzyme having one-third V_{max} and 10 times K_m of the other enzyme present. Inset is the graph displayed using a logarithmic scale for the substrate concentration.

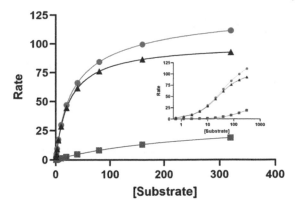

Failure to determine the initial rate of a reaction may also result in deviations from Michaelis–Menten kinetics. If an assay is used which, for example, determines the extent of reaction after a fixed time, it is possible that the reaction may have reached equilibrium, or completion during this time, especially at low substrate concentrations. This would result in the initial rates at low substrate concentrations being underestimated, and the curve appearing to be sigmoidal. Failure to measure initial rates may also hinder the ability of the assay to detect inhibition, as the uninhibited control rate slows down when there is depletion of substrate. It is therefore important to verify that the true initial rate is estimated at all concentrations of substrate used. This may be done by ensuring that the progress curve of the reaction, at all substrate concentrations, is linear, or less rigorously, by ensuring that only a small proportion of the substrate is converted to product over the measurement time.

5.6.2.2   Multiple Substrates

Most enzyme reactions are not of the simple one-substrate type as described previously. Instead, they involve two or more substrates, often resulting in the formation of more than one product. These enzyme-catalyzed reactions are classified by the number of substrates taking part in the reaction, the number of products formed and the order in which the substrates bind and the products released. The terms used to describe the order of substrate binding and product release are described below.

Sequential describes a system in which all the substrates bind to the enzyme before any product is released, see Reaction scheme 5.4. Sequential systems may be ordered if there is a required order for substrate addition.

$$E \; \rightleftharpoons \; \overset{A}{} \; EA \; \rightleftharpoons \; \overset{B}{} \; EAB \; \longrightarrow \; Products$$

Reaction scheme 5.4   Ordered sequential mechanism.

Random sequential mechanisms have no such requirement for substrate binding order, but still the substrates bind before release of any product, see Reaction scheme 5.5.

The Theorell-Chance mechanism (Reaction scheme 5.6) is an ordered mechanism, but the ternary complex does not accumulate.

Ping–pong describes a system in which one or more products are released between substrate additions (see Reaction scheme 5.7 in the Cleland notation) and involves reversible covalent modification of the enzyme.

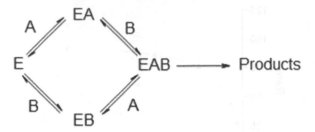

Reaction scheme 5.5 **Random ordered sequential mechanism.**

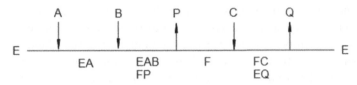

Reaction scheme 5.6 **Theorell-Chance ordered mechanism.**

Reaction scheme 5.7 **Cleland nomenclature for the bi uni uni uni ping–pong ter bi mechanism.**

There are specific terms used to describe the number of substrates and products involved in the reaction: uni, bi, ter and quad describe both the number of substrates and products involved in the reaction as shown below (Table 5.3).

For ping–pong systems, the individual substrate addition and product release steps may be denoted by the above nomenclature (Table 5.3). For example, in the ter bi system above, if A and B bind to the enzyme with release of P, before C binds and finally release of Q, this may be described as a bi uni uni uni ping–pong ter bi system.

A simplified notation, devised by Cleland can be used to describe the reaction sequences, for example, for the bi uni uni uni ping–pong ter bi system shown below (Reaction scheme 5.7).

It is important to note that the K_m for a particular substrate at one fixed set of conditions of the other substrates may not be the "real" K_m, but an apparent value that may change according to those conditions. Also, V_{max} observed at saturating levels of one substrate can be quite different

Table 5.3 Reaction nomenclature.

Reaction	Nomenclature
$A \rightarrow P$	uni uni
$A + B \rightarrow P$	bi uni
$A + B \rightarrow P + Q$	bi bi
$A + B + C \rightarrow P + Q$	ter bi
$A + B + C + D \rightarrow P + Q + R$	quad ter

to that under saturating conditions of another. The true K_m for a substrate can be considered as that observed under saturating conditions of all the other substrates. The true V_{max} is observed when all substrates are at saturating levels.

In the same way as K_m, inhibition constants may also be affected by the concentrations of substrates, and so an observed K_i may not be a true inhibitor dissociation constant.

Consider the scheme below (Reaction scheme 5.8), it can be shown that for many 2-substrate mechanisms, especially rapid equilibrium systems, the general expression for the initial rate is shown in Equation 5.11.

$$
\begin{array}{ccc}
E + A & \overset{K_a}{\rightleftharpoons} & EA \\
+ & & + \\
B & & B \\
\big\updownarrow K_b & & \big\updownarrow K_m{}^B \\
EB + A & \overset{K_m{}^A}{\rightleftharpoons} & EAB \overset{k_{cat}}{\longrightarrow} E + P
\end{array}
$$

Reaction scheme 5.8   Random order rapid equilibrium.

$$
v = \cfrac{V_{max}}{\left(1 + \dfrac{K_m{}^A}{[A]} + \dfrac{K_m{}^B}{[B]} + \dfrac{K_a K_m{}^B}{[A][B]}\right)}
$$

Equation 5.11   General expression for the rate of a 2-substrate reaction.
$K_m{}^A$ is the K_m for A determined at saturating [B], $K_m{}^B$ is the K_m for B determined at saturating [A] and K_a is the dissociation constant for the binary EA complex.

Parameter values for this model may be estimated using a global fit to rates measured for increasing concentrations of substrate A, under different fixed concentrations of substrate B (or vice versa) as seen in Figure 5.12.

Figure 5.12   Substrate dependence curves for varying substrate A at fixed concentrations of substrate B. Parameter values can be determined using a global fit to Equation 5.8.

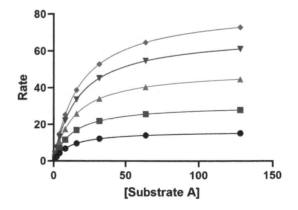

When the association of enzyme with one substrate has no effect on the binding of the second, then $K_a = K_m{}^A$, and $K_b = K_m{}^B$, so the relationship becomes Equation 5.12, below.

$$v = \frac{V_{max}}{\left(1 + \frac{K_a}{[A]} + \frac{K_b}{[B]} + \frac{K_a K_b}{[A][B]}\right)}$$

Equation 5.12   Expression for the rate of a 2-substrate mechanism, when the binding of one substrate has no effect on the binding of the other.

For the ping–pong bi bi mechanism, the value of K_a tends to zero, as the binding of A is effectively irreversible under initial rate conditions because P dissociates but is present at negligible concentrations and so cannot rebind. This leads to the modification of the rate equation (Equation 5.13)

$$v = \frac{V_{max}}{\left(1 + \frac{K_m{}^A}{[A]} + \frac{K_m{}^B}{[B]}\right)}$$

Equation 5.13   Expression for the rate of a ping–pong bi-bi mechanism.

For a bi bi compulsory ordered equilibrium mechanism (or for rapid binding of B) $K_m{}^A$ tends to zero, since saturating concentrations of B lead the concentration of free enzyme to tend to zero and so Equation 5.12 becomes Equation 5.14.

$$v = \frac{V_{max}}{\left(1 + \frac{K_m{}^B}{[B]} + \frac{K_a K_m{}^B}{[A][B]}\right)}$$

Equation 5.14   Expression for the rate of a bi bi compulsory ordered mechanism.

As mentioned above, reactions involving two (or more) substrates often follow Michaelis–Menten kinetics when one substrate is varied at a fixed concentration of the other(s). The calculated values of K_m and V_{max} often vary according to the concentration of the fixed substrate(s). The general equation can be rewritten as shown below (Equation 5.15)

$$v = \frac{\frac{V_{max}[A][B]}{(K_m{}^B + [B])}}{[A] + \left\{\frac{(K_a K_m{}^B + K_m{}^A[B])}{(K_m{}^B + [B])}\right\}} = \frac{V_{max}{'}[A]}{K_m{}^{A'} + [A]}$$

Equation 5.15   Alternative expression for the rate of reaction for a 2-substrate reaction.
Where the observed V_{max} is modified by a factor of $\frac{[B]}{(K_m{}^B + [B])}$ and the apparent $K_m{}^A$ is $\frac{(K_a K_m{}^B + K_m{}^A[B])}{(K_m{}^B + [B])}$.

Distinguishing between various different reaction mechanisms often requires further experiments than just initial velocity studies. These may need to take the form of product inhibition studies, binding studies, or isotope exchange methods.

5.6.3   Plate Type

Once an initial assay set-up is established, it is good practice to explore different plate types to understand any effects of the plate type on the target stability and its reaction behavior. This allows the choice of plate type to be made before the detailed optimization process is undertaken. Key considerations are the [E] versus rate relationship; understanding the range of enzyme concentrations and reaction times that can be used for further work.

5.7 Assay Optimization

Assay optimization attempts to address several key questions:

- What conditions are required for optimal enzyme activity?
- How do these conditions affect the detection system?
- What hit profile is expected from the screen?
- What compromises need to be made for optimal screen performance?

It is important to think about the optimization of both the detection system and the target enzyme system and ideally, both would be improved through the process. It is important to monitor each independently through the optimization process, so that issues may be identified early, and an attempt can be made to optimize one without compromising the other. It should be recognized that the perfect assay cannot be achieved, but it is critical to understand where compromises are being made.

Buffer identity, concentration and composition can have significant effects on enzymatic activities. It is also possible for buffer components to affect compound potency. Components of the assay to vary include divalent cations, for example, Mg^{2+}, Mn^{2+}, Ca^{2+}, etc.; inorganic salts, for example, NaCl and KCl; chelating reagents like EDTA; reducing agents such as DTT, TCEP, β-mercaptoethanol and glutathione; cofactors, for instance, NADH; detergents like Triton, CHAPS, Tween; DMSO concentration; buffer source, for example, HEPES versus acetate; pH; temperature; ionic strength. These components may often be varied in factorial experimental design (FED) experiments, see Section 5.7.1, which aim to identify optimal conditions with fewer experiments.

Considerations of buffer composition are critical not only for enzyme assays but also for maintaining enzyme integrity during long-term storage. Many enzymes must be stored at $-70\,°C$ to maintain activity, and repeated freeze–thaw cycles are not recommended. Other enzymes can be stored for long periods of time at $-20\,°C$ using additives in the storage buffer such as glycerol. Storage conditions should be checked by keeping a record of the enzyme activity and the potency of reference compounds over time.

The presence of carrier proteins in the buffer (e.g. bovine serum albumin, ovalbumin), although not recommended, as compounds may bind to these proteins, may also provide means of retaining enzyme activity. It is good practice to aliquot frozen reagents in appropriate sizes to minimize the freeze–thaw cycles as this can degrade reagents, in particular proteins. Calculating the amount of each reagent needed for the total assay size and aliquoting each reagent to allow this is good practice. This will help to reduce waste and will ensure that each aliquot will only have one freeze–thaw cycle.

5.7.1 Factorial Experimental Design (FED)

Various components of the buffer can be used as factors in FED experiments, which can be varied (in both identity and concentration) to optimize the enzyme assay. Literature data or previous experience can be useful in selecting factors to be included in FED experiments. For example, factors

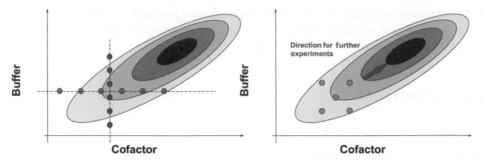

Figure 5.13 Comparison between one variable at a time (OVAT) optimization and factorial experimental design (FED).
In the OVAT example experiment (left), buffer concentration is varied at a fixed cofactor concentration and vice versa. In the FED example experiment, three different levels of each are tested simultaneously (right).

that could be considered for a FED experiment could be any of those listed above. The important difference between FED experiments and one variable at a time (OVAT) optimization experiments, is that the FED approach makes use of the ability to vary many factors simultaneously (Figure 5.13).

However, there are key experimental considerations for FED experiments that include using a fixed substrate concentration. There is no need to vary this, as it is often clear how the substrate concentration affects the measured rate, and this will be re-visited once the optimal conditions are identified. It is important to remember that conditions may affect the detection system as well as the target protein activity. Another consideration is that time and enzyme concentration are inter-dependent variables and so these values are also not usually varied during FED experiments. There is a clear correlation between these two variables and so, evaluating a correlation that is known to exist is not the best use of the experimental scope. The FED experiment allows the testing of many variables in one experiment, often at two or three concentrations. The aim is to evaluate the over-all experimental landscape, to identify interdependencies between variables and to rapidly identify regions of experimental space that allow for further optimization. It should be remembered that interdependencies between variables are both assay and target-dependent. These types of experiments provide an initial view of the robustness of the system, identifying variables that may be varied widely with little effect and those where tight regulation may be required for optimal performance. However, optimization of factors to produce the highest signal, when those factors have to be very carefully controlled, is perhaps not the overall goal. It is more useful to find a compromise whereby a reasonable signal can be obtained, which is resistant to small changes in variable concentration having a large effect on rate (Figure 5.14).

A narrow, steep optimum set of assay conditions may provide the highest rate but may lead to a steep decrease in rate if conditions are varied unexpectedly. A broad, shallow optimum will provide a more robust assay that will be resistant to large changes in rate if small changes in experimental conditions occur, as might be expected from day to day or between laboratory variation.

This will lead to robust assays that are more likely to deliver consistent parameter values, even if there are slight variations in experimental conditions. The number of factors that can change an assay performance is very high, and there is an almost infinite range of concentrations and combi-nations. The requirement for each enzyme is different and so the selection of factors to include in FED experiments must be empirically determined. There are several benefits of using FED experiments, including a better understanding of deconvoluting the effect of variables on the activity,

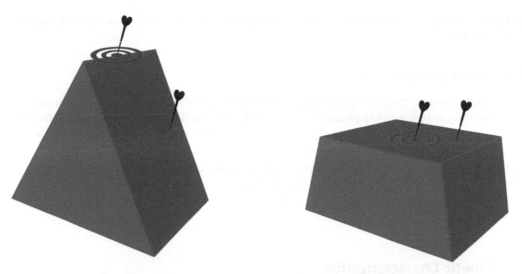

Figure 5.14 **Broad shallow optimum (right-hand side) versus a steep optimum (left-hand side).**
Demonstration that a broad, shallow optimum for assay conditions provides a stable environment, where variation in conditions will not significantly affect the measured rate. Conversely, a steep optimum may provide a higher rate, but if small variations in conditions lead to a sharp decrease in rate, then this optimum is not suitable for an assay design for screening over a long period of time, where changes in conditions may occur.

a better definition of the boundaries for valid data, reduction in the time taken for optimization, reduction in the use of materials and ability to easily combine with the use of automation.

Finally, artificial intelligence (AI) can be employed to assist in establishing FED approaches for enzyme assays, facilitating the optimization and efficiency of the experimental process. AI algorithms can analyze large datasets, explore various factors and identify optimal conditions to improve the accuracy and precision of enzyme assays. Employing AI methods allows exploration of the existing literature, databases, and experimental data to gather information on relevant factors affecting the potential assay. This knowledge can be used to build a foundation for designing factorial experiments and selecting appropriate factors to be included. AI also can help determine the levels or ranges of each factor to be tested in the FED experiment, based on historical data or simulation. This aids in establishing an effective experimental design that covers a wide range of conditions to obtain the most comprehensive information. Additionally, AI algorithms can employ optimization techniques, such as response surface methodology [13] or evolutionary algorithms [14], to determine the combination of factor levels that maximize enzyme activity or minimize variability. These algorithms can iteratively adjust factor levels and analyze the results to identify the optimal conditions for the assay.

Machine learning algorithms can also be utilized to model and predict enzyme assay outcomes based on the factorial experimental data collected. These models can help identify key factors influencing enzyme activity and predict optimal conditions for future experiments. AI algorithms also may suggest additional factors that may have been overlooked but are potentially influential, potentially expanding the scope of the experimental design.

AI's ability to analyze complex datasets, optimize factor combinations and predict outcomes offers the possibility for enhancing the design and execution of factorial experimental approaches during assay optimization, leading to improved accuracy and efficiency in designing effective assays.

5.7.2 Coupling Enzyme Considerations

If the detection system uses a coupled assay approach, it is useful to understand the requirements for the coupling enzymes before proceeding to the kinetic evaluation under optimized conditions. This requires knowledge of the threshold for the coupled systems commitment to product formation. The concentration of each enzyme required for the coupled process should be investigated to ensure that the excess is sufficient for system reliability. Other considerations include evaluating if there are additional complications arising from the nature of the coupled assay as well as the cost and availability of the coupling enzymes themselves, as often they will be required in much larger amounts than the target enzyme.

5.8 Kinetic Characterization

Kinetic characterization involves measuring the kinetic constants for the enzyme-catalyzed reaction under the conditions that will be used for screening. This allows an initial understanding of the most likely kinetic mechanism for the target, and an estimation of the distribution of enzyme species at different concentrations of substrate(s) to be derived. From this information, an appreciation of the bias the substrate concentration will place on the hit profile can be obtained and a decision made as to what substrate concentrations are best used.

5.8.1 Substrate Concentration

The choice of substrate concentration is important, but it is also important to realize that there may be no "correct" substrate concentration. There will always be advantages and disadvantages for the chosen concentration. The choice will be dictated by the assay performance versus the desired hit profile.

Often, the substrate(s) are used at their respective K_m values, so that there is a balance between the free enzyme and substrate-bound forms. This approach provides the minimum bias versus any mechanism of inhibition. However, bias may be introduced to enhance the probability of hit finding versus particular mechanisms if desired, by adjusting the substrate concentrations accordingly. For example, raising the substrate concentration well above K_m will bias the assay away from finding competitive compounds and will tend to favor uncompetitive compounds. Conversely, low substrate concentrations will favor competitive inhibitors and will bias away from uncompetitive compounds. Sometimes it may be advantageous to select physiological concentrations of substrate as this may help to understand the activity relative to cellular systems.

Clearly, understanding the behavior of the reaction with varying substrate concentration at this stage is a critical activity and these experiments are key to assay optimization as the substrate concentration impacts the rate, the reaction time and the signal. As they affect the distribution of enzyme species within the assay and inhibitors may bind with different affinities to these species, substrates are treated differently to other variables considered during assay optimization. Sometimes the relationship between substrate concentration and enzyme processing can be complex, but a detailed characterization at this stage is useful to understand what compromises can be made and what effect these may have for the outcome from compound screening.

Following the selection of the substrate concentration for running a primary compound screen, it is then possible to prepare a protocol highlighting the optimal conditions for the screen. This will comprise the enzyme concentration, substrate concentration, reaction time and the assay requirements in terms of additives and plate type. An understanding of the likelihood of finding hits versus the different mechanisms of inhibition will have been generated and a thorough assessment of the detection system and target biochemistry will have been undertaken. At this stage, the assay is ready to be assessed for its adaptability for transferring into the screening procedure and potential for automation.

5.9 Assay Adaptability

Assessing the assay adaptability involves understanding the assay requirements in terms of processes such as preincubation and reaction stop times, the number of reagent addition steps, the stability of reagents under screening conditions and the effect of temperature (even though it may be difficult to change this if using large-scale automation).

5.9.1 Tool Compounds

The selection of control compounds should also be undertaken to identify the options that may be suitable to provide a measure of 100% inhibition (so-called MIN control). Alongside this, activities to identify compounds that can be used to indicate 50% inhibition should be undertaken. This provides options for REF controls. It is prudent to assess whether the control compounds selected exhibit well-behaved concentration-response curves, how they behave under the plated batch conditions required for the screen and whether material is available to support their use as controls at the scale required.

5.9.2 DMSO Tolerance

Most compounds screened during early-stage drug discovery are kept as stock solutions dissolved in dimethyl sulfoxide (DMSO). As a result, there will be an amount of DMSO added to the assay, depending upon the stock concentration and the dilution factor to achieve the final concentration of the test compound used in the screen. These concentrations can often be very high in molar terms (100% DMSO = 14.1 M, so even 1% DMSO is 141 mM). Hence, the tolerance of the assay to DMSO must be evaluated to understand what effect these concentrations have on the assay. This allows the limitation around the range of DMSO concentration and hence the range of compound concentration that can be tolerated on the screen and subsequent follow-up concentration-response measurements.

5.9.3 Assay Stability

The stability and bulk storage requirements of the reagents must be assessed as should be the working stock reagent preparation procedures and incubation requirements. The reagents should be checked to ensure they are stable under the conditions for the experiment, for example, room temperature, under automation conditions (e.g. upon recirculation).

The requirement for stopping the reaction should also be assessed. This may be required to ensure that the reaction remains under steady-state conditions before the reaction is stopped and the signal measured, or it may be necessary for the automated process, to ensure that all plates are treated in the same way. For a quenched reaction, it is essential to test the stability of the stopped reaction over time to assess the behavior of the measured signal. For example, the background may increase or the signal decrease over time after the reaction has stopped, which may give erroneous estimates of the effect on compounds. It is equally important to check the behavior of the controls under the stopped assay conditions over time. There may be a requirement to stabilize the signal over time after the reaction has been stopped, by adjusting temperature or by including certain additives. This potentially allows the screen to be more flexible should there be unplanned disruptions or automation failures. Longer periods of stability will potentially allow the rescue of assay plates and reading of the signal should unforeseen events occur, minimizing the loss of data on a screening run. Understanding the range of reaction times and stability of the assay system over time allows an assessment of which conditions produce the same assay performance (for example Z') and provides an insight into whether the assay will allow extended detection applications, for example, the ability to read a plate again after a defined period of time.

5.9.4 Triage Assays

Alongside the primary assay that will be employed in screening for initial actives, the assays that will be used to confirm and triage those actives into real hits will also need to be considered. Attributes for these assays include the ability to filter out false positives or artifacts, the throughput required to enable testing of enough actives to sufficiently check the false positive rate. Often, alternative technologies are used in downstream assays as they provide an orthogonal assessment of the output, minimizing false positives due to compounds interfering with the detection technology. As with the primary assay, assays used in the hit triage cascade to understand specificity and selectivity should be optimized and potential issues that may arise on transfer to automated platforms identified.

5.10 Assay Evaluation (Validation)

It is recommended that all assays should have a plate uniformity assessment carried out. This study ideally should be run on two separate days and include three plates per day to assess uniformity and signal-to-noise. These plates should be tested at the DMSO concentration, which will be used in screening. The variability tests are then conducted on three types of signals: the max, min and mid signals.

The max signal (or neutral/total/top signal), labelled T in the plate layouts below, is the maximum signal measured in the assay. For enzyme assays, these would be the control values in the absence of an inhibitor, but with all other reagents included.

The minimum signal (or null/bottom signal), labelled B in the plate layouts below, measures the background signal. For enzyme assays, this represents the signal output from the assay at a maximally inhibiting concentration of a standard inhibitor, or the consensus minimum signal on the exclusion of one essential component of the assay at a time.

The mid signal (or reference signal), labelled M in the plate layouts below, estimates the signal variability at some point between the maximum and minimum signals. For enzyme assays, the mid-point is reached by adding an IC_{50} concentration of a known standard compound.

The following plate layouts, for which analysis templates can be developed may be used to measure assay variability (Figure 5.15). These plate layouts have allocated wells that produce

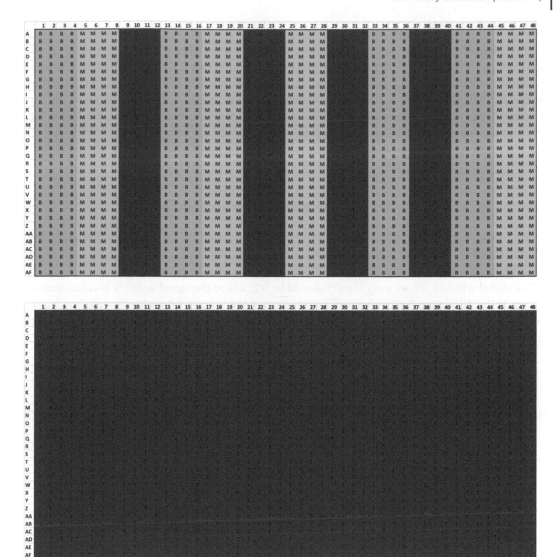

Figure 5.15 **Plate layouts for assessing assay variability.**

maximum, minimum and mid signals in a series of columns. It is recommended that for each type (make/material/supplier) of plate tested that two independent trials are run (on two separate days), using the desired assay protocol. These two experiments should use independently prepared reagents.

It is also possible to use single plates containing each of the maximum, mid and minimum conditions for a plate uniformity study, but this would require at least six plates per day. Experiments should then be performed using two full plates on each day for each experimental condition so that inter-plate variability can also be assessed. Homogeneous plate (plates each of maximum, mid and minimum signals) variability studies should be graphed using the data from all the plates in the experiment. Scatter plots from homogeneous plate layout studies can be interpreted in the same way as the plate layouts described.

5.10.1 Calculations

Useful calculations for plate uniformity studies are shown below. The mean, SD (standard deviation), and % CV for each signal (maximum, mid and minimum) on each plate should be calculated.

For every mid-signal well, a percent activity relative to the means of the maximum and minimum signals on that plate should be calculated as shown in Equation 5.16, where $\overline{\text{max}}$ and $\overline{\text{min}}$ are the average values for the maximum and minimum signals, respectively. The mean and SD for the mid-signal percent activity values on each plate should then be calculated.

$$\%\text{activity} = \frac{\text{mid} - \overline{\text{min}}}{\overline{\text{max}} - \overline{\text{min}}} \times 100$$

Equation 5.16   Expression for the percentage activity, calculated from controls.

Three other assay performance measures that can be calculated are the signal window (SW), the Z', and the assay variability ratio (AVR) which can all be calculated from the positive and negative controls (max. and min. data).

The signal window for an assay ideally should be >2, where the signal window is calculated as shown below (Equation 5.17), where $\overline{\text{max}}$ and $\overline{\text{min}}$ are the average values for the maximum and minimum signals, respectively, and SD_{max} and SD_{min} are the standard deviations of the maximum and minimum signals, respectively.

$$\text{SW} = \frac{(\overline{\text{max}} - 3\text{SD}_{\text{max}}) - (\overline{\text{min}} + 3\text{SD}_{\text{min}})}{\text{SD}_{\text{max}}}$$

Equation 5.17   Expression for the signal window (SW).

The Z'-factor is calculated as shown below in Equation 5.18, and for an ideal assay should be greater than 0.5, where the parameter values are the same as above.

$$Z' = \frac{(\overline{\text{max}} - 3\text{SD}_{\text{max}}) - (\overline{\text{min}} + 3\text{SD}_{\text{min}})}{(\overline{\text{max}} - \overline{\text{min}})}$$

Equation 5.18   Expression for the Z' factor.

The AVR is calculated as follows (Equation 5.19), and it is recommended that for a useful assay, the value is less than 0.6, where the parameter values are the same as above. Note that the $\text{AVR} = 1 - Z'$, so they will have the same statistical properties.

$$\text{AVR} = \frac{3(\text{SD}_{\text{max}} + \text{SD}_{\text{min}})}{(\overline{\text{max}} - \overline{\text{min}})}$$

Equation 5.19   Expression for the assay variability ratio (AVR).

It is reasonable to assume that the SD of the max signal is at least as large as the SD of the min signal. With this assumption, it is possible to plot the relationship between Z' and SW (Figure 5.16).

Simulated data [15] have shown that although the sampling properties of both Z' and SW show positive bias (tend to overestimate the true value for the assay), the bias for Z' was lower. The Z' also had a lower sampling variability than SW, as measured by the ordinary CV. Thus, it is recommended that Z' is used to measure assay performance because it shows better bias and precision.

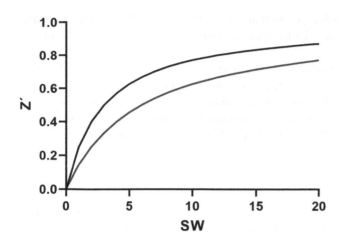

Figure 5.16   Relationship between *Z'* and SW.
The black line defines the limit of the relationship when $SD_{max} \gg SD_{min}$, and the red line defines the relationship when the $SD_{max} = SD_{min}$.

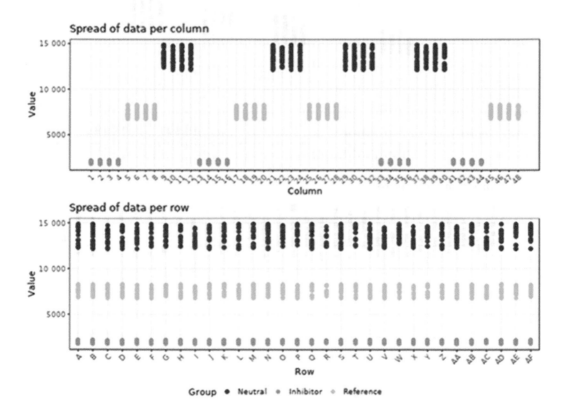

Figure 5.17   Plots of response versus well number for data indicating no drift or edge effects.
The purple circles indicate the maximum signals, the blue circles are the mid signals, and the yellow circles are the minimum signals.

Note that if individual plates for each experimental condition are used then, Z-factor, signal window, and mid-point activity must be calculated by aggregating the data across all the plates for each day's experiment.

Other useful measures of assay performance are the signal: noise (S:N) and the signal: background (S:B) ratios. S:N is defined as the mean signal (maximum or minimum signal) divided by the standard deviation of that signal. This can be considered as a measure of the signal "strength" of the assay. S:B is defined as the ratio between the mean maximum signal and the mean minimum or background signal. This can be useful for describing the dynamic range of the assay.

5.10.2 Assessing Plate Uniformity

Scatter plots can help to reveal patterns associated with problems in the assay. For example, drift (systematic variation in values across or down a plate), edge effects (where the outer wells give

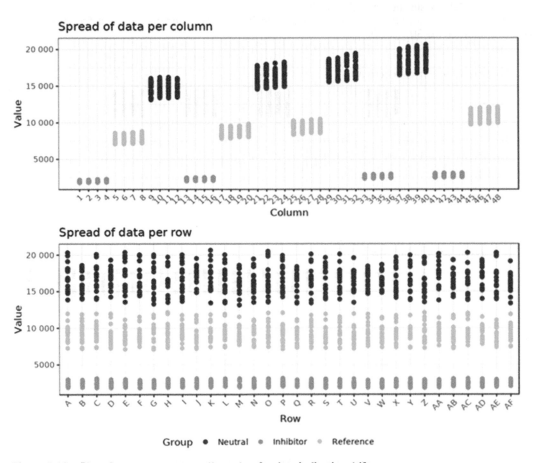

Figure 5.18 Plot of response versus well number for data indicating drift.
It can be seen that the signals from each set of control wells are increasing across the plate. The purple circles indicate the maximum signals, the blue circles are the mid signals, and the yellow circles are the minimum signals.

responses significantly lower than the inner wells) and other systematic causes of variability. The response is plotted against well number, where the wells are ordered either by row or column.

The following two plots (of the same data) are plotted in these different ways to show an example where there are no edge effects or drift, and the controls are well separated (Figure 5.17).

The maximum and mid signals can be used to investigate drift. Drift associated with the minimum signals should only be considered if the mean minimum signal is greater than 10% of the maximum signal. Significant trends in the signal from left-to-right and top-to-bottom are usually fairly easy to observe. Drift that exceeds 20% should be investigated. In the example below (Figure 5.18), the drift in response to the mean maximum values across the plate is 26%, which is significant and should be investigated.

Edge effects can also contribute to variability and recognizing them can be valuable. Edge effects can sometimes be due to evaporation from wells that are incubated for long periods of time. Edge effects can also be caused either by short incubation times or by plate stacking. These conditions may allow the edge wells to reach the desired incubation temperature faster than the inner wells. These types of effects may appear as shown in the graphs below (Figure 5.19).

Data showing systematic trends can be even more apparent when the data are visualized as a heat map (Figure 5.20). In this representation, the data are displayed in the layout of the microtiter plate and the intensity of the response is represented using a color scale, as shown below.

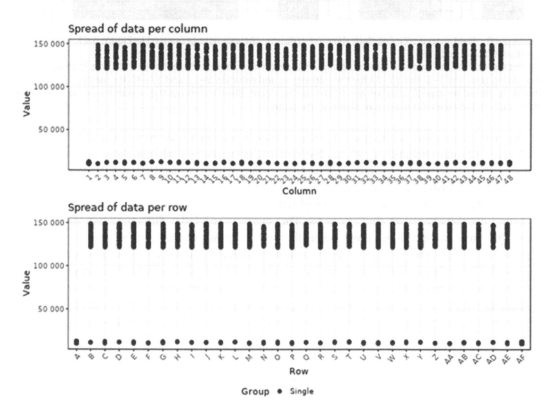

Figure 5.19 **Plots of response versus well number for data indicating a strong edge effect – decrease in signal.**
The purple circles indicate the maximum signals.

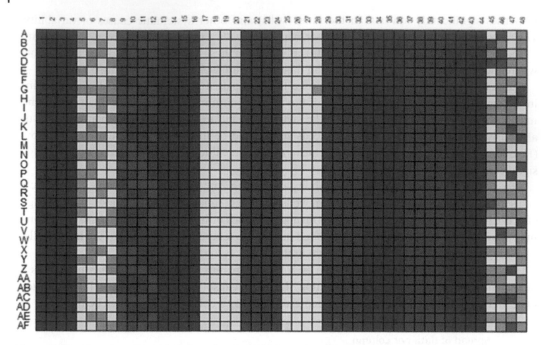

Figure 5.20 **Heat map displaying data showing drift from the example presented in Figure 5.17, using the top layout of Figure 5.14.**
It can be seen that the mid and maximum signals are increasing across the plate, by 26%, this is shown by comparing columns 5–8 and 45–48 for the mid signal, and columns 9–12 and 37–40 for the max signal. This drift should be investigated further as 26% is significant. Blue represents the minimum signal in the plate and red indicates the maximum signal in the plate.

This heat map is a contour map displayed as octiles, where each color represents 12 wells, with the highest 12 values being dark red, and the lowest 12 values being dark blue. Intermediate bands of 12 wells are represented by intermediate colors. Heat maps can also be constructed around % effect or other statistical calculations.

Ideally, the normalized mid signal should not show any significant shift across assay plates or between experiments on separate days. The significance of the difference will depend to a certain extent on the slopes observed in the dose–response curves, and so plate-to-plate or experiment-to-experiment variation in the mid-point percent activity needs to be examined in light of the slope factor of the dose–response curves of the assay. Typically, for enzyme assays, the slope factor is fixed at 1, so a 20% difference in signal will correspond to a 1.5-fold change in potency. Ideally, there should not be a shift in potency of greater than 1.5 between any two plates within an experiment, or greater than 2 between any two different average experimental mid-point % activities.

5.10.3 Acceptable Assay Performance Criteria

An acceptable assay should have intra-plate variability criteria that fall within the ranges shown in Table 5.4 for each individual plate. It should have no significant observable edge or other spatial effects.

Ensuring that assay statistics fall within these ranges provides a greater likelihood of success for identifying true actives during the screening run.

Table 5.4 Table of acceptable criteria for assay performance to begin screening.

Assay statistical parameter	Acceptable criteria
CV_{max} and CV_{mid}	$\leq 20\%$
Normalized SD_{mid}	$\leq 20\%$
Normalized SD_{min}	$\leq min$
Z'	>0.5
SW	>6

References

1 Bilban M, Billich A, Auer M, Nussbaumer P. New fluorogenic substrate for the first continuous steroid sulfatase assay. *Bioorganic & Medicinal Chemistry Letters*. 2000;10(9):967–9.

2 Zhong W, Benkovic SJ. Development of an internally quenched fluorescent substrate for *Escherichia coli* leader peptidase. *Analytical Biochemistry*. 1998;255(1):66–73.

3 Nishikata M, Yoshimura Y, Deyama Y, Suzuki K. Continuous assay of protein tyrosine phosphatases based on fluorescence resonance energy transfer. *Biochimie*. 2006;88(7):879–86.

4 King MJ, Sharma RK. *N*-myristoyl transferase assay using phosphocellulose paper binding. *Analytical Biochemistry*. 1991;199(2):149–53.

5 Holt A. Radiochemical assay of monoamine oxidase activity. *Methods in Molecular Biology (Clifton, NJ)*. 2023;2558:45–61.

6 Nishikata M, Suzuki K, Yoshimura Y, Deyama Y, Matsumoto A. A phosphotyrosine-containing quenched fluorogenic peptide as a novel substrate for protein tyrosine phosphatases. *Biochemical Journal*. 1999;343(Pt 2):385–91.

7 Suárez AS, Stefan A, Lemma S, Conte E, Hochkoeppler A. Continuous enzyme-coupled assay of phosphate- or pyrophosphate-releasing enzymes. *BioTechniques*. 2012;53(2):99–103.

8 Atienza J, Tkachyova I, Tropak M, Fan X, Schulze A. Fluorometric coupled enzyme assay for *N*-sulfotransferase activity of *N*-deacetylase/*N*-sulfotransferase (NDST). *Glycobiology*. 2021;31(9):1093–101.

9 Henri V. Théorie générale de l'action de quelques diastases. *Comptes rendus de l'Académie des Sciences*. 1902;135:916–9.

10 Michaelis L, Menten ML. Die kinetik der invertinwirkung. *Biochemische Zeitschrift*. 1913;49:333–369.

11 Johnson KA, Goody RS. The original Michaelis constant: translation of the 1913 Michaelis–Menten Paper. *Biochemistry*. 2011;50(39):8264–9.

12 Yang J, Copeland RA, Lai Z. Defining balanced conditions for inhibitor screening assays that target bisubstrate enzymes. *Journal of Biomolecular Screening*. 2009;14(2):111–20.

13 Onyeogaziri FC, Papaneophytou C. A general guide for the optimization of enzyme assay conditions using the design of experiments approach. *SLAS Discovery*. 2019;24(5):587–96.

14 Pettersen JP, Almaas E. Parameter inference for enzyme and temperature constrained genome-scale models. *Scientific Reports*. 2023;13(1):6079.

15 Iversen PW, Eastwood BJ, Sittampalam GS, Cox KL. A comparison of assay performance measures in screening assays: signal window, Z' factor, and assay variability ratio. *SLAS Discovery*. 2006;11(3):247–52.

6

Enzyme Inhibition

6.1 Introduction

Inhibitors are molecules that are capable of binding to an enzyme and preventing or reducing its normal activity. The presence of an inhibitor in an enzyme assay decreases the rate of the enzyme-catalyzed reaction by binding to the enzyme and preventing it from converting substrate to product. Inhibitors may also function by binding to the substrate and preventing its turnover to product, but this behavior will not be discussed here. Inhibitors can completely or partially inhibit the enzyme in a concentration-dependent manner, where higher concentrations of inhibitor result in higher levels of inhibition until maximum inhibition is reached, when the rate of reaction drops to zero for a full inhibitor, or to a defined background rate for a partial inhibitor. The rate of reaction in the absence of the inhibitor is the uninhibited reaction rate, which is usually used as a control in all assays.

The measured rate is a measure of uninhibited enzyme concentration. This can be easily seen since in the absence of an inhibitor the rate is equal to the control rate (all the enzyme is uninhibited), and when the enzyme is fully bound by the inhibitor, the rate will drop to zero, as no turnover occurs (all the enzyme is inhibited). Studying enzyme inhibitors usually requires product versus time to be linear for steady-state conditions to apply and rate versus enzyme concentration to be linear to be sensitive to inhibition. A linear relationship then means that a drop of 50% in the

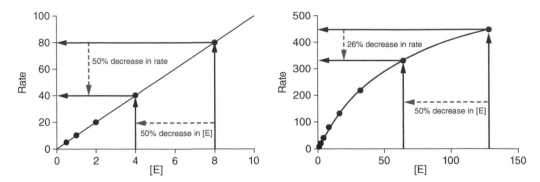

Figure 6.1 **Dependence of rate on enzyme concentration.**
A typical linear (left) or nonlinear (right) plot showing the dependency of rate on enzyme concentration [E]. At low [E], there is a linear relationship with [E] and so a 50% decrease in [E] results in a 50% decrease in rate. At higher [E], the relationship between [E] and rate is nonlinear. In this plot, a 50% decrease in [E] only results in a 26% decrease in rate.

concentration of uninhibited enzyme concentration would lead to a corresponding 50% drop-in rate (Figure 6.1).

This graph shows a typical plot of the dependence of rate on the enzyme concentration in the assay. Several possible reasons for the curvature have been discussed above. The graph clearly shows that for a linear dependence of rate versus enzyme concentration (left hand side graph), and to obtain reliable parameters for inhibition studies, enzyme concentrations greater than around 10 (arbitrary units) cannot be used. The use of low enzyme concentrations is also desirable for other reasons, which will be discussed below.

The degree of inhibition is then measured using concentration-response analysis by determining the rate of reaction at different concentrations of inhibitor, where the concentration of active enzyme (E′, with no inhibitor bound) is proportional to the rate (see below).

Note that most enzyme kinetic experiments study the steady-state (see Chapter 5), when the concentration of intermediates is approximately constant with respect to time. The parameters measured in such experiments are apparent values, which may be composed of contributions from several different physical steps. These constants are therefore macroscopic constants. This contrasts with measurements made during the pre-steady state, where the concentration of intermediates changes with respect to time and the aim is to measure absolute parameters, which relate to individual physical steps in the reaction mechanism. These values are microscopic constants. The value of a parameter under a specific set of experimental conditions at a particular time is termed an apparent value and is denoted with the prime symbol (′).

6.2 Substrate and Product Inhibition

High concentrations of substrate can sometimes lead to significant inhibition of enzyme activity. Similarly, high concentrations of the product may also inhibit enzyme action. Substrate inhibition is described in more detail in Chapter 5 but can often be avoided by working at lower substrate concentrations [1].

Key Example: Substrate Inhibition of Phosphofructokinase

Phosphofructokinase is an important regulatory enzyme in glycolysis. It catalyzes the phosphorylation of fructose 6-phosphate (F-6-P) to fructose 1,6-bisphosphate (F-1,6-P$_2$) using ATP. It is inhibited by high levels of its own substrates (fructose 6-phosphate and ATP). The inhibition by ATP occurs via an allosteric mechanism and may be overcome by increased concentration of AMP, which functions as an allosteric activator [1].

Product inhibition can often be avoided by measuring initial rates so that only a small amount of product is generated during the assay, although it should be remembered that those products binding tightly may still inhibit even at low concentrations [2]. It is important to realize that in the presence of product inhibition, the net rate in the presence of product is always smaller than the initial rate which is estimated in the absence of product. This is because (i) at any time some of the product is being converted back to substrate (for reversible reactions, even at equilibrium, substrate is converted to product and product to substrate, but at equilibrium the forward and reverse rates are the same), and (ii) some of the enzyme is bound to product, therefore lowering the concentration available to bind to substrate.

Key Example: Product Inhibition of Hexokinase

In another example from glycolysis, hexokinase catalyzes the transfer of the gamma phosphoryl group from ATP to the oxygen at C6 of glucose, to form glucose-6-phosphate and ADP. It is allosterically inhibited by physiological concentrations of glucose-6-phosphate. This mechanism helps to control the rate of utilization of glucose [2].

6.3 Reversibility

The concept of reversibility is not absolute in the study of enzyme kinetics, but depends upon the timescale of the assay. For example, the drug Darunavir, which inhibits the HIV-1 protease, has a dissociation half-life of greater than 10 days, illustrating that although the inhibition is eventually reversible, observations on the timescale of a few minutes in a traditional enzyme assay would appear to show irreversible behavior for this compound.

6.3.1 Testing for Irreversible Inhibition

The general approach to test for reversible inhibition is to incubate the enzyme and inhibitor together and then try to remove the inhibitor and allow the catalytic activity to be regained. There may be many ways to attempt removal, including large dilution (jump dilution), dialysis or gel filtration. Inhibitors that dissociate slowly may appear irreversible and so methods to increase the apparent rate of dissociation of the enzyme-inhibitor complex may be employed to help in this situation, for example, by including a molecule that competes for the inhibitor binding site, such as the substrate. It can also be helpful in the identification of irreversible inhibitors to add additional substrate and additional enzyme to the reaction, as shown in Figure 6.2.

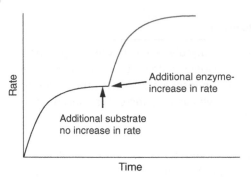

Figure 6.2 **Test for irreversible inhibition.** Once the reaction in the presence of an irreversible inhibitor has stopped (all the enzyme is complexed with inhibitor) addition of extra substrate will not increase the rate of reaction because the enzyme remains irreversibly bound to the inhibitor. However, the addition of further enzyme will increase the rate initially but will follow the same trend as the initial reaction, if the inhibitor is in excess.

It is important to add that irreversible inhibitors are not required to form covalent bonds with the enzyme and that a compound that does form a covalent bond with the enzyme may do so reversibly. Thus, the distinction between reversible and irreversible binding is made on kinetic grounds. Compounds that fall between rapidly reversible and irreversible are misleadingly termed slow-binding inhibitors. These compounds may associate rapidly with the enzyme and often the rate-limiting step is a slow isomerization step, which leads to the formation of a higher affinity complex. The dissociation of the inhibitor can also be slow. Irreversible inhibition can therefore be considered as a special case of slow-binding inhibition which will be later described (Section 6.7).

Inhibition is usually assumed to be reversible, unless suggested by a poor fit for the concentration–response equation, known reactivity of the inhibitors, or precedence for the enzyme under study.

6.3.2 Rapidly Reversible Inhibition

Several assumptions are usually made when studying the kinetics of rapidly reversible inhibition:

1. The system is at steady-state, so that E, S and I are all in equilibrium.
2. The total inhibitor concentration, $[I]_t$ is much greater than the total enzyme concentration, $[E]_t$, so that the free inhibitor concentration $\approx$ total inhibitor concentration, $[I]_f = [I]_t$. Stated another way, there is an insignificant amount of inhibitor bound to the enzyme.
3. The total substrate concentration, $[S]_t$ is much greater than the total concentration of enzyme, so that there is also an insignificant amount of substrate bound to the enzyme, and so $[S]_t = [S]_f$.

Often these conditions are valid for test compounds, and these are therefore known as classical inhibitors. However, the assumptions may break down leading to slow-binding inhibition and a type of inhibition known as tight-binding inhibition, see Table 6.1.

Slow-binding and tight-binding are especially common in drug discovery where the aim is to find potent inhibitors. $[E]_t$ may then be similar to $[I]_t$ and the onset of inhibition may be slow because

Table 6.1 **Different types of reversible inhibitors.**

Type of inhibition	Concentration of free inhibitor required	Time for equilibration between E and EI
Classical	$\gg [E]_t$	$\ll$assay time
Slow binding	$\gg [E]_t$	Not $\ll$ assay time
Tight binding	Around or below $[E]_t$	$\ll$assay time
Slow, tight binding	Around or below $[E]_t$	Not $\ll$assay time

it is dependent on the low concentration of inhibitor. These types of inhibition will be discussed below (Sections 6.6 and 6.7).

6.4 The IC$_{50}$ Value

The degree of inhibition caused by compounds, which inhibit enzyme reactions, is often characterized by IC$_{50}$ values. The IC$_{50}$ is the concentration of a compound that inhibits enzyme activity by 50%, compared to the uninhibited control rate. The IC$_{50}$ of a compound, when the assumptions above hold, is effectively the same as K_i' and so represents its potency and the smaller the value the more potent the inhibitor. An IC$_{50}$ for a compound can be calculated by measuring enzyme activity at different compound concentrations (see Chapter 5). The value of IC$_{50}$ may change when assay parameters, such as preincubation, substrate concentration or other factors are altered.

6.4.1 Determining the IC$_{50}$ Value

Where the IC$_{50}$ value of a compound is unknown, several experiments may need to be completed to optimize the compound concentration range to provide an accurate IC$_{50}$ value (Protocol 6.1). A wide concentration range, with many data points, covering the possible range of IC$_{50}$ values, often from micromolar to nanomolar may be initially required. Ideally, these concentrations should be geometrically spaced (using a constant dilution factor). This usually facilitates experimentation as a constant dilution simplifies the preparation of inhibitor concentrations via serial dilution (Figure 6.3).

Once an estimate of IC$_{50}$ is obtained, a concentration range centered around the IC$_{50}$ value covering at least 0.2–5.0 (and ideally 0.1–10) × IC$_{50}$ is usually adequate (this provides a range of inhibited rates from 83.3% of the uninhibited rate to 16.7% of the uninhibited rate, or ideally 90.9% to 9.1%).

6.4.2 Use of pIC$_{50}$

The potency of an inhibitor often is reported as a pIC$_{50}$ value. The pIC$_{50}$ value represents the negative logarithm of the IC$_{50}$ in molar (Equation 6.1).

$$pIC_{50} = -\log[IC_{50}](M)$$

Equation 6.1 Expression showing the relationship between pIC$_{50}$ and IC$_{50}$.

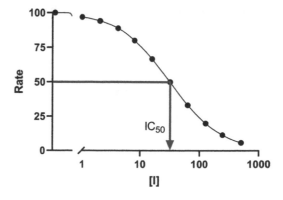

Figure 6.3 Optimized determination of IC$_{50}$. An example of good experimental design for inhibitor concentration–response studies. Data points are collected at geometrically spaced concentrations (2-fold dilutions) covering a wide range of concentrations. The IC$_{50}$ is indicated as the concentration of inhibitor (assuming all assumptions are valid) giving 50% of the control rate. The log x-axis is useful to display the wide range of inhibitor concentrations, the lower concentrations of which would be bunched together using a linear plot.

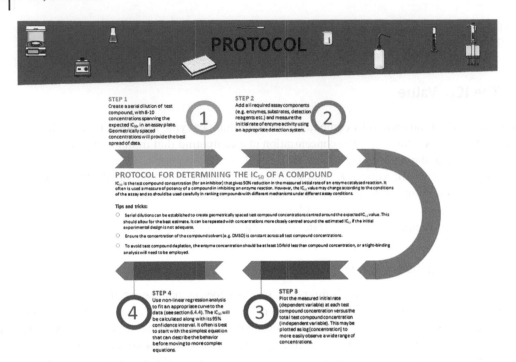

Protocol 6.1   How to determine IC$_{50}$.

There are several advantages of using pIC$_{50}$ instead of IC$_{50}$. Firstly, using the log scale means the numbers are more manageable. This is comparable to the use of the pH scale which is very familiar to scientists and successfully manages the large range of hydrogen ion concentrations experienced in biological systems. Similarly, an IC$_{50}$ of 1 µM is 10^{-6} M, which is pIC$_{50}$ = 6.0, IC$_{50}$ of 100 nM is 10^{-7} M, which is pIC$_{50}$ = 7.0, an IC$_{50}$ of 10 nM is 10^{-8} M, which is pIC$_{50}$ = 8.0, an IC$_{50}$ of 1 nM is 10^{-9} M and which is pIC$_{50}$ = 9.0. Therefore, higher values of pIC$_{50}$ indicate more potent compounds. Comparisons can easily be recognized, and it is simple to see that a 10-fold change in IC$_{50}$ is observed as a 1-unit change in pIC$_{50}$. The use of pIC$_{50}$ also removes the potential issue for confusion with units. For example, it is clear that a pIC$_{50}$ of 3 represents an IC$_{50}$ of 1 mM, and a pIC$_{50}$ of 6 represents and IC$_{50}$ of 1 µM. However, the units would be critically important for conferring the information about the 1000-fold difference in potency if just the values of 1 were reported. The use of pIC$_{50}$ encourages the use of logarithmic thinking and so prevents the use of zero or negative values for concentrations. It also means that taking average potency values is facilitated, since the arithmetic mean of pIC$_{50}$ values can be used, rather than having to calculate geometric means of IC$_{50}$ values.

6.4.3   Comparison of Potency

IC$_{50}$ values are often used as a measure of the potency for enzyme inhibitors, but these values should be used with caution, because, as mentioned above, IC$_{50}$ values are not constants, and their values are specific to the experimental conditions during the assay. For example, IC$_{50}$ values may change with the enzyme preparation and concentration, substrate identity and concentration, the concentration of cofactors, coenzymes or other factors. Hence, IC$_{50}$ values frequently change when there are changes in assay conditions. Thus, there are certain considerations that should be taken when comparing potencies using IC$_{50}$ values:

IC$_{50}$ values should be corrected in order to remove effects from tight-binding kinetics; sufficient time must be allowed for steady-state to be reached before measuring rate; it is important to assay all inhibitors under identical conditions when comparing IC$_{50}$ values against a single target; it is best to assay enzymes under conditions as close to physiological as possible, using the substrate of importance in the disease state; the same source of enzyme and purification procedure should be used.

Thus, the IC$_{50}$ should be considered only as a qualitative estimate of potency since the measured values are not absolute. Microscopic equilibrium constants are required to fully evaluate apparent dissociation constants. These may be obtained through direct binding measurements to the relevant enzyme forms or via pre-steady state kinetics, which often requires large amounts of purified enzyme. Careful comparison of IC$_{50}$ values thus, is often the only available method for assessing potency.

6.4.4 Concentration-response Curve Analysis

For classical inhibitors, it can be shown that the rate in the presence of inhibitor (v_i) can be related to the rate in the absence of inhibitor (v_0), using an expression containing the apparent dissociation constant for the inhibitor, K'_i and the free inhibitor concentration, [I] (Equation 6.2). Consider the scheme below (Reaction scheme 6.1).

$$E' + I \xrightleftharpoons{K'_i} EI$$

Reaction scheme 6.1 Reaction scheme showing reversible inhibitor binding.

Where E' represents the enzyme not bound by I. For classical kinetics (i.e. not tight binding), we can assume that $[I]_t \gg [I]_{bound}$, so that $[I]_t = [I]_f$ and that $IC_{50} = K'_i$. Rate in the absence of I is $k_{cat}[E]_t$; so that $v_0 = k_{cat}[E]_t$. Rate in the presence of I is $k_{cat}[E']$, where [E'] is enzyme not bound by I; so that $v_i = k_{cat}[E']$.

$$K'_i = \frac{[E'][I]}{[EI]}$$

By conservation of mass

$$[E]_t = [E'] + [EI]$$

Substituting for [EI] from above,

$$[E]_t = [E'] + \left(\frac{[E'][I]}{K'_i} \right)$$

Rearranging,

$$[E]_t = [E'] \left(1 + \frac{[I]}{K'_i} \right)$$

and

$$\frac{v_i}{v_0} = \frac{[E']}{[E]_t} = \frac{1}{\left(1 + \frac{[I]}{K'_i} \right)}$$

so

$$v_i = \frac{v_0}{\left(1 + \frac{[I]}{K_i'}\right)}$$

when

$$v_i = \frac{v_0}{2}, [I] = K_i' = IC_{50}$$

Equation 6.2   Expression for the 2-parameter concentration-response equation.

This equation is derived based on an appropriate model and is valid for all modes of inhibition but, the values of K_i' relative to K_i are different for different modes of inhibition (Chapter 8). This should be the first model used in data analysis, with more complex models applied only if appropriate.

There are other equations that can also be fitted to the data for enzyme inhibition, which have only a single additional parameter compared to the simple 2-parameter concentration response equation shown above (Equation 6.2). For example, simple + offset (background), where the rate does not reduce to zero (Equation 6.3).

$$v_i = \frac{v_0}{1 + \frac{[I]}{K_i'}} + bg$$

Equation 6.3   Expression for a 3-parameter concentration-response equation (addition of offset from the simple 2-parameter equation).

Another equation with three parameters, where the slope is not equal to 1 is shown below (Equation 6.4).

$$v_i = \frac{v_0}{1 + \left(\frac{[I]}{IC_{50}}\right)^h}$$

Equation 6.4   Expression for a 3-parameter concentration-response equation (addition of a slope factor).

Additionally, a 4-parameter equation accounting for biphasic inhibition, which describes the effect of added inhibitor on rate, for instance, if there are two enzyme species (e.g. phosphorylated and non-phosphorylated) inhibited to different extents by the added inhibitor (Equation 6.5).

$$v_i = \frac{v_{01}}{1 + \frac{[I]}{K_{i_1}'}} + \frac{v_{02}}{1 + \frac{[I]}{K_{i_2}'}}$$

Equation 6.5   Expression for a 4-parameter concentration-response equation accounting for biphasic inhibition.

More commonly, an equation known as the 4-parameter logistic (Equation 6.6) is fitted to concentration-response data. It contains two extra parameters compared to the simple 2-parameter equation that can describe inhibition, and so should be used with caution. As can be seen from the derivation of the simple concentration response equation, inhibitor binding with $1:1$ stoichiometry should have a Hill slope of 1, and zero background (Equation 6.2).

The variables in the 4-parameter logistic are v_0, IC$_{50}$, slope factor (h) and background (bg). These parameters are often justified for cell data, or *in vivo* data, but are not usually justified for isolated protein (purified enzyme) data. Fitting this equation can mask important characteristics, such as tight binding, slow binding or multiple phases.

$$v_i = \frac{v_0}{1 + \left(\frac{[I]}{IC_{50}}\right)^h} + bg$$

Equation 6.6 **Expression for the 4-parameter logistic equation.**

Observation of slope factors that are different from 1 may be indicative of more complex behavior, as the slope factor may reflect stoichiometry of interaction (number of molecules of I bound per molecule of E). Usually, this is equal to 1. However, slope factors above 1 may indicate tight binding, aggregation or positive cooperativity (binding of 1 inhibitor molecule favors binding of another). Slope factors below 1 may signify multiple phases which can be due to slow binding, or negative cooperativity (binding of 1 molecule disfavors binding of another).

Often, in drug discovery, the 4-parameter logistic equation is used as it can be used to analyze data from several different experimental situations and is often implemented into software for automated data fitting. It is useful to evaluate the results of fitting a 4-parameter logistic by considering the behavior of the data and the resultant fitted curve.

In addition, it can be useful to replace the parameter K_i' with $10^{\log K_i'}$ (compare with pIC$_{50}$, above) when fitting these equations to the generated data. This does not change the value of the fitted parameter, but it has been shown that these values conform more closely to a normal distribution when converted to logarithms [3]. It also ensures a positive number for the parameter with the units of concentration and leads to realistic 95% confidence intervals for that dataset.

6.4.4.1 Bell-shaped Behavior

Sometimes the highest inhibitor concentrations give rates that are not as low as expected for the concentrations used, relative to the expected IC$_{50}$ and lower concentrations. This results in curves that display a bell-shape. Sometimes, this is due to compound insolubility or aggregation, resulting in a lower concentration of compound than expected. If there is a strong reason to exclude these data points, then they may be excluded from the curve-fit, which may help improve the confidence in the estimated IC$_{50}$. It may be possible to use alternative solvents to improve solubility, but caution should then be used in comparing IC$_{50}$ values (Figure 6.4).

Figure 6.4 **Curve illustrating bell-shaped behavior.**
Testing high concentrations of inhibitor ([I]) can sometimes result in bell-shaped behavior, this may occur when there is aggregation or insolubility. Bell-shaped behavior can hinder the accurate determination of IC$_{50}$. Caution should be taken when comparing the IC$_{50}$ for these types of curves.

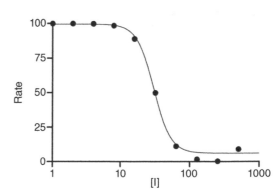

6.4.4.2 Weakly Active Compounds

A weak compound may not exert a full response over the tested concentration range giving rise to curves that do not reach a full plateau (Figure 6.5). There is lower confidence in calculated IC_{50} values where there are few concentrations above the IC_{50}. Increasing the compound concentration can help to provide better estimates, but this may be technically prohibitive (e.g. the stock concentration of the compound may not high enough or this may result in a higher solvent concentration than is tolerated by the assay).

6.4.4.3 Steep Curves

Often compounds identified in high-throughput screening campaigns demonstrate steep slopes in concentration responses, where the Hill slope is >2 (Figure 6.6). In the absence of tight binding, this may be indicative of compound-induced protein aggregation and/or super-stoichiometry. Often, these compounds may not be suitable for further optimization, but if more thorough estimation of the IC_{50} is required since there are usually few points on the sloping part of the curve, a further experiment with more concentrations around the mid-point may provide a better estimate of IC_{50}.

6.4.4.4 Partial Curves

Screening hits also can produce concentration-response curves that have a clear point of inflection, but which do not lead to full inhibition (Figure 6.7). Such partial responses suggest that the compound, although demonstrating concentration-dependent inhibition is, for some reason, not able to fully inhibit the enzyme reaction. This suggests that the enzyme is still able to turnover the product in the presence of saturating concentrations of the inhibitor. Usually, partial responses do not affect the determination of IC_{50}, but it is useful to distinguish these partial inhibitors from full inhibitors, especially when selecting compounds for further chemical optimization.

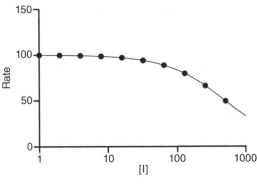

Figure 6.5 Curve illustrating weakly active compounds.
Low-potency inhibitors can result in an incomplete curve over the tested concentration range, this means the IC_{50} of these curves cannot be determined accurately. Caution should be taken when comparing the IC_{50} of these curves as the confidence interval surrounding the estimated IC_{50} will be large.

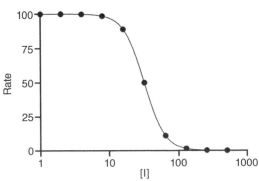

Figure 6.6 Curve illustrating steep concentration response.
Steep concentration-response curves are defined as those with a Hill slope > 2, these types of curves can show "all or nothing" responses as there are occasions where there are <2 points on the slope of the curve and therefore the calculated IC_{50} can be inaccurate. Steep curves can indicate super-stoichiometry and/or compound-induced protein aggregation, so caution should be taken when progressing compounds with this behavior.

Figure 6.7 Curve illustrating partial concentration response.
Partial curves are defined as curves that show a full curve across the concentration of inhibitor ([I]) tested (i.e. there are defined upper and lower plateaus of the curve) but do not show 100% inhibition and the inhibitor is incapable or fully inhibiting the enzyme even at high [I]. In this case, the determination of IC_{50} is not affected but it is useful to identify this type of behavior as this may be less desirable than inhibitors that inhibit by 100%.

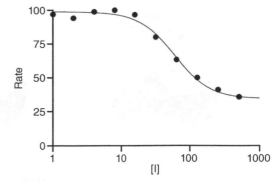

Figure 6.8 Curve illustrating noisy concentration response.
Sometimes during hit identification, the concentration curve can be unoptimized and noisy, which means determination of an accurate IC_{50} can be challenging. In noisy concentration-response curves, the confidence in the calculated IC_{50} can be low, increasing replicates can increase the confidence in the IC_{50}.

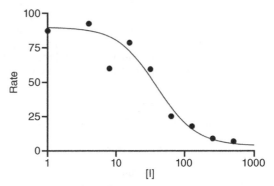

6.4.4.5 Noisy Data

The limitations of high-throughput assays, combined with compounds from screening libraries, may sometimes lead to "noisy" concentration responses, where the data points follow a trend but are often distant to the best-fit line (Figure 6.8). Often, to improve the confidence in the estimated IC_{50}, an increased number of replicates will be required.

6.5 Identity of Substrate

The identity of the substrate may affect the observed inhibition. Model substrates may often be used because it is easier to follow the reaction using these simpler substrates. Model substrates are commonly smaller than the physiological substrate. Examples include peptides as models for proteins (proteases, kinases), p-nitrophenol phosphate for phosphatases and esters for proteases. The use of these smaller substrates may preclude the detection of inhibitors that bind adjacent to the catalytic site. For example, some hirudin analogues bind at an "exosite" on thrombin (Figure 6.9). They inhibit activation of fibrinogen, but not turnover of peptide-coumarin [4].

Alternatively, inhibitors may not be detected if different enzyme forms predominate for the natural substrate. For example, catalysis by serine proteases involves the formation of an acyl-enzyme intermediate. Amides (and proteins) tend to be rate-limited by the formation of the acyl-enzyme. However, esters tend to be rate-limited by the hydrolysis of acyl enzyme [5, 6] (Reaction scheme 6.2). Thus, an assay using an ester substrate favors inhibitors, which bind to the acyl enzyme.

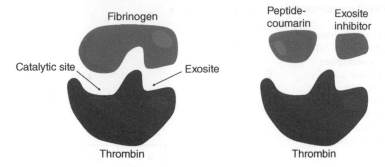

Figure 6.9 Using model substrates in drug discovery.
Fibrinogen binds to two sites on thrombin, the catalytic site and the subsite. In biochemical assays when targeting thrombin, a coumarin peptide may be used instead of the physiological substrate fibrinogen, as this is a simpler assay setup. However, inhibitors have been identified that inhibit the activation of fibrinogen but do not inhibit the turnover of the simple coumarin peptide. This is because the smaller peptide only occupies the catalytic site, and the inhibitor binds to the subsite. This illustrates a potential issue when using simple, model substrates as mimics of full, physiological substrates.

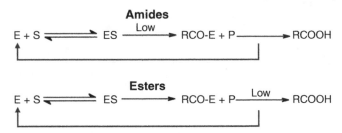

Reaction scheme 6.2 Reaction scheme illustrating serine protease catalysis for different substrates.

6.6 Effect of Enzyme Concentration – Tight-binding Inhibition

When the concentration of free inhibitor required is around or below the $[E]_t$, then assumption 2 above (Section 6.3.2) breaks down and the situation of tight-binding inhibition occurs.

From Reaction scheme 6.1 above, Equation 6.2 also holds for tight-binding compounds, where the free $[I]$ required to give significant inhibition is not $\gg[E]_t$. Thus, a significant amount of $[I]_t$ is depleted by formation of the EI complex. In this case, $[I]_f$ does not approximate to $[I]_t$, so that Equation 6.2 becomes Equation 6.7.

$$K_i' = [E'] \frac{([I]_t - [EI])}{[EI]}$$

The observed rate, v_i is 50% of v_0 when $[E'] = [EI]$, so that

$$K_i' = [I]_t - \frac{[E]_t}{2}$$

Equation 6.7 Expression showing the relationship between K_i' and enzyme concentration under tight-binding conditions.

The relationship between IC_{50} and K_i' can be defined by rearrangement as shown in Equation 6.8.

$$IC_{50} = K_i' + \frac{[E]_t}{2}$$

Equation 6.8 **Expression demonstrating the relationship between K_i' and IC_{50}.**

This relationship is valid for both classical and tight-binding inhibition. In classical inhibition, the enzyme concentration is much lower than the inhibitor concentration and so the equation simplifies to $IC_{50} \approx K_i'$. In tight-binding situations where the enzyme concentration is not negligible compared to the inhibitor concentration and the inhibitor is potent ($K_i' \sim 0$) then $IC_{50} \approx [E]_t/2$. In this case, the measured IC_{50} value no longer reflects the potency of the compound and is limited by the enzyme concentration. The IC_{50} value will equal half the enzyme concentration until the enzyme concentration is sufficiently low enough to calculate the inhibitors true potency.

For tight-binding inhibitors, the relationship describing the inhibitor dependence of rate, regardless of mechanism of inhibition is shown in Equation 6.9.

$$v_i = v_0 \left\{ -0.5 \left(\frac{K_i'}{[E']} + \frac{[I]}{[E']} - 1 \right) + 0.5 \sqrt{ \left[\left(\frac{K_i'}{[E']} + \frac{[I]}{[E']} - 1 \right)^2 + \frac{4K_i'}{[E']} \right] } \right\}$$

Equation 6.9 **Expression for tight-binding inhibition.**

The contribution of $[E]_t/2$ is significant for tight-binding inhibitors, which demonstrates that the observed value of IC_{50} does not accurately measure potency for any compound that follows tight-binding kinetics. When K_i' is not $\gg[E]_t/2$ nor $\ll[E]_t/2$, the observed IC_{50} underestimates potency.

In the extreme case, where K_i' is $\ll[E]_t/2$, the observed IC_{50} is independent of potency and is equal to $[E]_t/2$, regardless of the absolute value of K_i'. The determination of K_i' for these very tight-binding compounds is very difficult because only a small fraction of the added inhibitor is free in solution.

An advantage of tight-binding inhibition is that the concentration of functional enzymes (or more precisely the concentration of binding sites) can also be estimated using tight-binding inhibitors. In this approach, experiments are established in which [E] and [I] are much greater than K_i' (values of [E] at least 10-fold higher than K_i' are recommended, in the example below [E] = $5 \times K_i'$). Under these conditions, nearly all the added inhibitor becomes bound to the enzyme and the rate falls almost linearly with increasing inhibitor concentrations until the entire enzyme is bound with inhibitor (at [E] = [I]) and the rate approaches zero. At higher inhibitor concentrations, the rate will remain close to zero. The enzyme concentration is calculated from the tight-binding equation or stoichiometric-binding equation or estimated from the extrapolation to zero rate.

Determination of concentration of the binding sites using a tight-binding inhibitor is shown below (Figure 6.10). The value of [E] ([E] = 10 in this example) is calculated using the tight-binding equation or may be estimated ([E] $\approx$ 11.1 in this example) by the intersection of the extrapolated line with the x-axis. The higher the ratio of [E]: K_i', the more closely the extrapolated line will match the true value of [E].

An alternative method of calculating the concentration of binding sites is to vary the enzyme concentration at several known, fixed inhibitor concentrations (Figure 6.11).

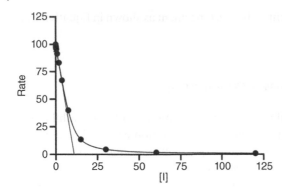

Figure 6.10 Use of a tight-binding inhibitor to estimate enzyme concentration.
Nearly all the added inhibitor is bound at concentrations below the enzyme concentration. At higher concentrations, the enzyme is saturated, and the rate tends toward zero. The extrapolation (red line) of the initial linear portion gives an estimate of enzyme concentration at the intersection on the *x*-axis.

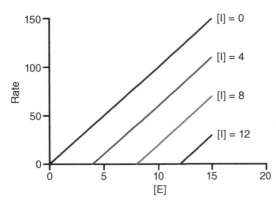

Figure 6.11 Alternative method for using a tight-binding inhibitor to estimate enzyme concentration.
Here, the enzyme concentration is varied at different fixed concentrations of a tight-binding inhibitor. The rate only increases above a measurable level once the enzyme concentration has increased above the concentration of the added inhibitor. The intercepts give a measure of the active enzyme concentration, relative to the total added.

6.7 Slow-binding Inhibition

When the time for equilibration between E and EI is not much shorter than the assay time, then assumption 1 (Section 6.3.2) above breaks down and the situation of slow-binding inhibition occurs (Reaction scheme 6.3). A 1-step slow-binding inhibition mechanism results in the linear relationship between the observed rate constant (k_{obs}) and [I] shown in Figure 6.16.

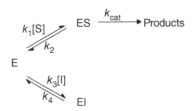

Reaction scheme 6.3 Reaction scheme illustrating a 1-step slow-binding inhibition mechanism.

When an inhibitor has a low K_i' value and [I] is varied in the region of K_i', values for the pseudo-first-order rate constant for the formation of EI ($k_3 \cdot$ [I]) and also for the dissociation rate constant (k_4) would be low. These low values of association and dissociation would lead to slow binding even though k_3 may be of the order expected for a diffusion-controlled reaction. It is the product of the true second-order rate constant (k_3), the inhibitor concentration ($[I]_t$) and the enzyme concentration ([E]) that determines the rate of EI formation. It can also be seen that if

[E] is very low, possibly because the majority of $[E]_t$ is actually in the form ES, due to very high competing substrate, that slow-binding may result.

Slow-binding inhibition may also occur because of a low value of k_3, which may occur due to barriers that the inhibitor encounters in its binding at the active site of the enzyme. Slow-binding inhibition can also arise if there is a rapid formation of an initial collision complex (EI), which subsequently undergoes a slow isomerization to a higher affinity complex (EI*). A reaction scheme for this type of slow-binding inhibition is shown below (Reaction scheme 6.4). This type of mechanism is a 2-step mechanism where the relationship between the rate (k_{obs}) and [I] is hyperbolic as shown in Figure 6.17.

Reaction scheme 6.4   **Reaction scheme for a 2-step slow-binding inhibition mechanism.**

In this case, the overall dissociation constant (K_i^*) would be defined by Equation 6.10.

$$K_i^* = \frac{[E][I]}{[EI] + [EI^*]} = \frac{K_i k_6}{k_5 + k_6}$$

where

$$K_i = \frac{k_4}{k_3}$$

Equation 6.10   **Expression for K_i for a 2-step slow-binding inhibition mechanism.**

The extent to which K_i^* is lower than K_i will depend on the relative magnitudes of the values for k_5 and k_6. For example, if $k_6 \ll k_5$, $K_i^* \ll K_i$, and the equilibrium will favor EI* formation. If $k_5 \ll k_6$, $K_i^* \approx K_i$, formation of EI* is insignificant and classical competitive inhibition will be observed. Hence, for inhibition to conform to the mechanism above (Reaction scheme 6.4), K_i^* must be lower than K_i, and so k_6 must be lower than k_5. Also, the values of k_5 and k_6 must be of a magnitude that allows observation of the attainment of equilibrium between EI and EI*.

There is a great deal of evidence to suggest that the majority of slow-binding and slow, tight-binding inhibitions occur according to a 2-step mechanism [7]. It is envisaged that inhibitors initially combine with enzymes at their active sites, and subsequently induce conformational changes that cause an increase in affinity and the formation of a more stable enzyme-inhibitor complex, from which inhibitor is released slowly.

6.7.1   Progress Curves for Slow-binding Inhibition

Progress curves for slow-binding inhibitors show curvature over time when the uninhibited control rate is linear.

Each inhibited curve for the 1-step mechanism (Reaction scheme 6.3) contains an initial segment, the slope of which does not vary with inhibitor concentration and is identical in magnitude to that in the absence of inhibitor (i.e. the control rate). This shows that the initial rate (v_0) of slow-binding (or irreversible) inhibitors, conforming to a 1-step mechanism, being invariant with inhibitor concentration, gives no information relating to potency (Figure 6.12). It can be seen (more easily for higher inhibitor concentrations) that after this initial portion of the progress curve, the rate decreases over time until a final steady-state rate is reached. An essential characteristic of any slow-binding inhibition plot is that the curves at higher concentrations of inhibitor exhibit asymptotes (Figure 6.13), representing these final steady-state rates. The observation of these asymptotes indicates that the decrease in reaction rate does not occur because of enzyme inactivation, decrease in substrate concentration or product inhibition. It is usually difficult to observe steady-state rates at lower inhibitor concentrations, since the reaction rate is too fast, relative to the time taken to obtain significant concentrations of EI, and the decrease in substrate concentration becomes significant.

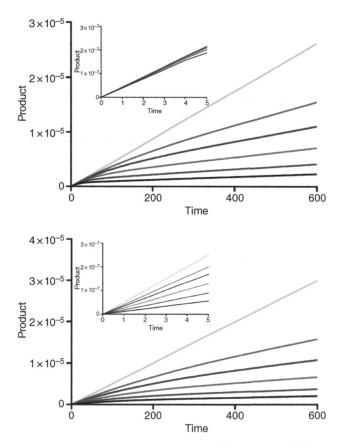

Figure 6.12 Progress curves for 1-step (top) and 2-step (bottom) slow-binding inhibition.
The top plot shows a progress curve of a 1-step slow-binding inhibition mechanism where concentration of inhibitor ([I]) is varied. In the presence of inhibitor ([I] $\neq$ 0), there is visible curvature in the progress curve, whereas the control rate ([I] = 0) is linear. Initial rates taken from earlier time points show that there is no difference in rate in the presence or absence of the inhibitor (inset). The bottom plot shows a progress curve of a 2-step slow-binding inhibition mechanism. As in the 1-step mechanism, the presence of inhibitor shows visible curvature in the progress curve and the control rate remains linear. Initial rates taken from earlier time points show that there is a difference in rate in the presence of increased inhibitor (inset).

Figure 6.13 Time-course for 1-step slow-binding mechanisms highlighting the asymptotes which represent the steady state rates for each inhibitor concentration.

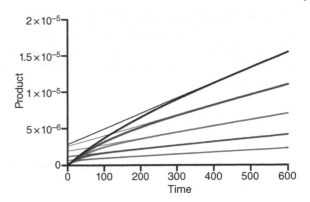

The progress curves for slow-binding inhibition are described by the general integrated equation (Equation 6.11) where v_0, v_s and k represent the initial velocity, the final steady-state velocity and the apparent first-order rate constant for the establishment of equilibrium between free enzyme and free inhibitor and EI* complex (or can be described as the apparent rate constant for v_0 changing to v_s).

$$[P] = v_s t + \frac{(v_0 - v_s)(1 - e^{-kt})}{k}$$

Equation 6.11 Expression for the general integrated equation for slow-binding inhibition.

For a 1-step mechanism (Reaction scheme 6.3), v_0 is invariant with inhibitor concentration, but the variation of v_s for a competitive inhibitor is described by the general equation for classical competitive inhibition (Equation 6.12)

$$v_s = \frac{V_{max}[S]}{K_m \left(1 + \frac{[I]}{K_i}\right) + [S]}$$

Equation 6.12 Expression for steady-state rate (v_s) for a competitive 1-step slow-binding inhibition mechanism.

V_{max} is the maximum velocity ($= k_{cat} \cdot [E]_t$), [S] is concentration of substrate, K_m is the Michaelis–Menten constant, and K_i is the dissociation constant for EI* complex ($= k_4/k_3$), which determines potency.

It can be seen from the progress curves that as the concentration of inhibitor increases, so does the rate at which the progress curve changes from v_0 to v_s. This is a function of the apparent first-order rate constant (k). The value of k is given by Equation 6.13 where k increases in a linear fashion with [I].

$$k = k_4 + \frac{k_3[I]}{\left(1 + \frac{[S]}{K_m}\right)}$$

Equation 6.13 Expression for the apparent first-order rate constant, k, for a competitive 1-step slow-binding inhibition mechanism.

It can be difficult to see from the progress curves for a 2-step mechanism (Reaction scheme 6.4) that there is an initial segment of each curve whose length and slope decreases as the concentration of inhibitor increases (Figure 6.12). The variation of initial rate (v_0) with [I] for a competitive inhibitor is given by Equation 6.14 where the symbols have the same meaning as above (see Equation 6.12).

$$v_0 = \frac{V_{max}[S]}{K_m \left(1 + \frac{[I]}{K_i}\right) + [S]}$$

Equation 6.14   Expression for initial rate, v_0, for a competitive 2-step slow-binding inhibition mechanism.

As with slow-binding inhibition following a 1-step mechanism (Reaction scheme 6.3), progress curves for competitive inhibitors conforming to a 2-step mechanism (Reaction scheme 6.4), also exhibit asymptotes representing steady-state rates. The variation of steady-state rate with inhibitor concentration is described by the equation for classical competitive inhibition (Equation 6.15), where the symbols are as described above (see Equation 6.12), except for K_i^*, which represents the overall dissociation constant.

$$v_s = \frac{V_{max}[S]}{K_m \left(1 + \frac{[I]}{K_i^*}\right) + [S]}$$

Equation 6.15   Expression for steady-state rate, v_s, for a competitive 2-step slow-binding inhibition mechanism.

The apparent first-order rate constant, k, for inhibitors conforming to a 2-step mechanism (Reaction scheme 6.4), described in Equation 6.16 increases hyperbolically with increasing inhibitor concentration, at a fixed concentration of substrate. The value of k is described by the following equation.

$$k = k_6 + k_5 \left\{ \frac{[I]}{K_i \left(1 + \frac{[S]}{K_m}\right) + [I]} \right\}$$

Equation 6.16   Expression for the apparent first-order rate constant, k, for a competitive 2-step slow-binding inhibition mechanism.

It should be noted that for inhibitors following this type of mechanism, that potency is determined by both the values of K_i, and K_i^*.

It can be seen from Table 6.2. that the half-times for inhibitor dissociation can range from minutes to days for slow-binding inhibitors.

Above, the progress curves for the reactions of enzymes with slow-binding inhibitors have been described for reactions started by the addition of enzymes. It is also possible to preincubate the enzyme and inhibitor and then start the enzyme reaction by the addition of substrate. In this case, the progress curve will take a different shape, being upward concave (Figure 6.14). Equation 6.11 will still hold, but now the value of v_0 will be lower than v_s, as the inhibitor is displaced by substrate and the rate of reaction increases during the assay.

Table 6.2   Slow-binding inhibitors of some enzymes, with their half-times for inhibitor dissociation.

Enzyme	Inhibitor	$t_{1/2}$ for activity regain	Reference
Chymotrypsin	MeoSucc-Ala-Ala-Pro-BoroPhe-OH	12 mins	Kettner and Shervi [14]
Angiotensin converting enzyme	Captopril	12 mins	Shapiro and Riordan [15]
Cathepsin G	Leupeptin	12 mins	Baici et al. [16]
HMG-CoA reductase	Compactin	15 mins	Nakamura and Abeles [17]
Dihydrofolate reductase	NADPH-methotrexate	27 mins	Stone and Morrison [18]
ACP enoyl reductase	Triclosan	1 hour	Ward et al. [8]
Squalene synthase	3-(Biphenyl-4-yl)-quinuclidine	1 hour	Ward et al. [19]
Pepsin	Pepstatin	2.5 hours	Rich and Sun [20]
Chymotrypsin	Isatoic anhydride	1 day	Moorman and Abeles [21]
Chymotrypsin	Aza peptides	5 days	Gupton et al. [22]
DNA polymerase	acyclovir triphosphate	42 days	Furman et al. [23]

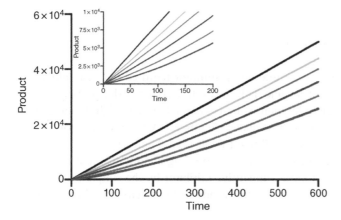

Figure 6.14   Progress curves of slow-binding inhibition when reaction initiated with substrate.
Preincubation of slow-binding inhibitors with the enzyme prior to reaction initiation with substrate shows upward curvature in their progress curves. In the absence of the inhibitor, the reaction is linear and as the inhibitor concentration ([I]) increases the observed upward curvature increases.

For the 1-step mechanism (Reaction scheme 6.3), the initial velocity for a competitive slow-binding inhibitor when the reaction is started with substrate is given by Equation 6.17.

$$v_r = \frac{V_{max}[S]}{(K_m + [S])\left(1 + \frac{\alpha[I]}{K_i}\right)}$$

Equation 6.17   Expression for initial velocity, v_r, for a competitive 1-step slow-binding inhibition mechanism where the reaction is initiated with substrate.

v_r is the initial velocity for the reaction started with substrate and α is the ratio of the final assay volume to the volume of the preincubation. Here, it is assumed that equilibration of the enzyme and inhibitor has occurred during the preincubation. This may be difficult to ascertain but allowing for several half-lives should be sufficient. The steady-state rate will, of course, be the same as that for the reaction started with the enzyme, where the concentrations are the same.

The corresponding experiment also may be carried out for competitive compounds conforming to the 2-step mechanism (Reaction scheme 6.4). In this case, the initial velocity is given by Equation 6.18.

$$v_r = \frac{V_{max}[S]}{K_m\left(1 + \frac{[I]}{K_i}\right) + [S]} \left\{ \frac{1 + \frac{\alpha[I]}{K_i}}{1 + \frac{\alpha[I]}{K_i^*}} \right\}$$

Equation 6.18 **Expression for initial velocity, v_r, for a competitive 2-step slow-binding inhibition mechanism where the reaction is initiated with substrate.**

To use the above equation for the analysis of slow-binding inhibition, it is essential that the progress curve for the reaction in the absence of inhibitor is linear, so that any decrease in rate is due to the action of the inhibitor alone. These equations assume that there is no significant substrate depletion.

6.8 Slow, Tight-binding Inhibition

When the value of K_i^* is very small, the concentration of inhibitor required to cause inhibition may become similar to the concentration of enzyme. This means that [I] can no longer be assumed to be equal to $[I]_t$. When this occurs, the inhibition is then classified as slow, tight binding. The progress curves will have the same shapes as those above (Figure 6.12), but the variation of initial velocity with inhibitor concentration may be difficult to observe. The variation of v_0 may be detected at higher inhibitor concentrations, where the tight-binding condition would not apply (since, $[I] \approx [I]_t$) and would be described for a competitive inhibitor by the simple competitive relationship shown above (Equation 6.12). However, the steady-state velocity would be given by Equation 6.19.

$$v_s = \frac{k_{cat}[S]Q}{2(K_m + [S])}$$

where

$$Q = \sqrt{\left[\left(K_i' + [I]_t - [E]_t\right)^2 + 4K_i'[E]_t\right]} - \left(K_i' + [I]_t - [E]_t\right)$$

and

$$K_i' = K_i^*\left(1 + \frac{[S]}{K_m}\right)$$

Equation 6.19 **Expression for steady-state rate, v_s, for competitive slow tight-binding inhibition.**

The equations above describe competitive tight-binding inhibition under steady-state conditions and allow for the change in concentration of inhibitor as a result of the formation of EI.

It is not possible to derive an explicit integrated equation to describe this type of inhibition, and so a value of k cannot be given. However, it is possible to analyze data for slow, tight-binding inhibition, provided that the difference in K_i and K_i^* is sufficiently great, so that the steady-state concentration of EI is negligible. Hence, the inhibition mechanism would then be described by a 1-step slow-binding mechanism (Reaction scheme 6.3). For this mechanism, an integrated equation has been derived see Equation 6.20.

$$[P] = v_s t + \frac{(v_0 - v_s)(1 - \gamma)}{k\gamma} \ln\left(\frac{1 - \gamma e^{-kt}}{1 - \gamma}\right)$$

where

$$\gamma = \frac{K_i' + [E]_t + [I]_t - Q}{K_i' + [E]_t + [I]_t + Q}$$

and

$$k = \frac{k_{cat} K_m Q}{K_m + [S]}$$

Equation 6.20 Expression for the integrated equation for 1-step slow tight-binding inhibition. Where definitions of v_0, v_s and K_i are given above (Equations 6.11 and 6.12).

The value of the dissociation constant (K_i) for a tight-binding inhibitor can be obtained by fitting the initial velocity data, obtained as a result of varying the total enzyme concentration at different fixed total concentrations of inhibitor, and a single fixed concentration of substrate to a modified form of Equation 6.19 giving Equation 6.21.

$$v = \frac{k_{cat}[S]R}{2(K_m + [S])}$$

where

$$R = \sqrt{\left[\left(K_i' + [I]_t\right)^2 + 2\left(K_i' - [I]_t\right)[E]_t + [E]_t^2\right]} - \left(K_i' + [I]_t - [E]_t\right)$$

So, when $[I]_t = 0$, v is given by:

$$v = \frac{k_{cat}[S][E]_t}{(K_m + [S])}$$

The true K_i value is obtained using the relationship:

$$K_i = \frac{K_i'}{\left(1 + \frac{[S]}{K_m}\right)}$$

Equation 6.21 Expression for K_i for a competitive slow tight-binding inhibitor.

It is interesting that reliable estimates for the kinetic parameters associated with a slow, tight-binding inhibitor can be obtained by fitting progress curves to Equations 6.14 and 6.16 provided that:

$$2[E]_t \leq [I]_t \geq K_i^* \left(1 + \frac{[S]}{K_m}\right)$$

For analysis of tight-binding inhibition, it is essential to know accurately the total concentration of enzyme used in the assay. This can be measured by an active site titration, or by measuring the initial velocity of the reaction after preincubation of the enzyme with varying concentrations of inhibitor [8, 9].

Key Example: Slow, Tight-binding Inhibitors of Enoyl-Acyl Carrier Protein Reductase

Triclosan is a widely used antibacterial agent, present in mouthwashes, toothpastes and dermatological products. Its antibacterial activity results from binding to the enzyme enoyl (acyl carrier protein) reductase. Triclosan binds slowly and weakly to the free enzyme but binds more rapidly and potently to the enzyme NAD^+ product complex. Its kinetic and structural characteristics have been studied versus the enzyme from *Escherichia coli* [8] and several slow, tight-binding inhibitors have been designed for the enzyme from *Mycobacterium tuberculosis* [9].

6.8.1 Conditions where Detection of Slow-binding may be Precluded

It can be seen, that if the rate of catalysis is high compared to the rate of isomerization of the EI complex, substrate depletion could occur before the steady-state equilibrium between EI and EI* is established. Under these conditions, a slow-binding inhibitor would appear to behave as a classical inhibitor. To demonstrate the slow-binding behavior, it is necessary to reduce the enzyme concentration, so that the equilibrium between EI and EI* is established before significant substrate depletion occurs. This demonstrates the need for collecting progress curve data at different enzyme concentrations in the study of slow-binding inhibitors.

Another reason for failure to detect slow-binding may be the termination of the reaction in the presence of an inhibitor before it has proceeded to the extent over which the reaction is linear in the absence of inhibitor. This demonstrates the value of collecting time-course data, even if eventually only initial rates are used.

Another reason for the failure to detect slow-binding is preincubation of inhibitor and enzyme, when nonlinear progress curves are observed on starting the inhibited reaction with enzyme, and determination of only initial rates.

6.9 Irreversible Inhibition

When analyzing progress curves (Figure 6.15), it can be seen that it may be very difficult to distinguish between a reversible inhibitor and one which is considered to be irreversible over the time

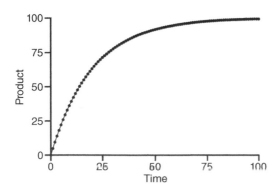

Figure 6.15 Progress curve showing irreversible inhibition.
The progress curve for an irreversible inhibitor can be difficult to distinguish from a reversible inhibitor progress curve.

Figure 6.16 Progress curves for an irreversible 2-step mechanism.
Progress curves for an irreversible 2-step inhibitor. In the absence of inhibitor, product formation is linear with time. As the inhibitor concentration increases, there is more curvature in the progress curves and the curves approach a plateau quicker.

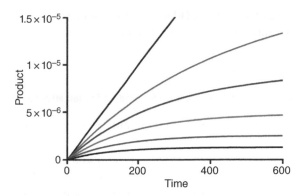

of the assay. The difference is that the steady-state rate of a reversible, slow-binding inhibitor will have a finite value (Figure 6.12), whereas the steady-state rate for an irreversible inhibitor is zero (as can be seen from the asymptotes at the higher inhibitor concentrations in Figure 6.16). It is therefore advisable to carry out further experiments, and to employ different techniques to verify the reversible, or otherwise, nature of the compound under study.

The equations describing the irreversible inhibition shown in the graph above can all be derived from those for slow-binding inhibition, by recognizing that both k_4 and the steady-state rate (v_s) are zero, for the 1-step slow-binding inhibition mechanism, Equation 6.11 now becomes the following equation (Equation 6.22).

$$[P] = \frac{v_0(1 - e^{-kt})}{k} = [P]_\infty(1 - e^{-kt})$$

Equation 6.22 Expression for the progress curve for an irreversible inhibitor.

The second part of the equation above is introduced since, when t tends to zero, $[P]$ tends to zero, but when t tends to ∞, $[P]$ reaches a constant value, $[P]_\infty$ $(= v_0/k)$, and k is now defined in Equation 6.23, which represents the amount of substrate converted to product before all the available enzyme has become bound as EI* complex.

$$k = \frac{k_3[I]}{\left(1 + \frac{[S]}{K_m}\right)}$$

Equation 6.23 Expression for the apparent rate constant, k, for a one-step, competitive and irreversible inhibitor.

For irreversible inhibition, the potency of the inhibitor is determined by the value of k_3 since K_i is not defined. The concentration of substrate turned over to product before inactivation of the enzyme by the irreversible inhibitor is given by the following equation (Equation 6.24).

$$[P]_\infty = \frac{V_{max}[S]}{k_3[I]K_m}$$

Equation 6.24 Expression for the amount of product formed at infinite time for a one-step, competitive and irreversible inhibitor.

which is obtained by substituting Equations 6.23 and 5.4 (the Michaelis–Menten equation) into the expression given in Equation 6.25. This arises as, when t tends to zero, $[P]$ tends to zero, but when t tends to ∞, [P] reaches a constant value, $[P]_\infty$ given by Equation 6.25, and from inspection of Equation 6.22 above, when t tends to ∞.

$$[P]_\infty = \frac{v_0}{k}$$

Equation 6.25   Expression for product formed after an infinite time period for a one-step mechanism.

For irreversible inhibition, conforming to a 2-step inhibition, where k_6 is zero time-courses are shown in Figure 6.16.

As for irreversible inhibition, conforming to a 1-step mechanism, product accumulation in the presence of competitive irreversible inhibitors following a 2-step mechanism (Reaction scheme 6.4) is described by Equation 6.22, however, k is now described by Equation 6.26.

$$k = \frac{k_5[I]}{K_i \left(1 + \frac{[S]}{K_m}\right) + [I]}$$

Equation 6.26   Expression for the value of the observed rate constant for decrease in rate over time for a competitive irreversible inhibitor conforming to a 2-step mechanism.

It can also be shown, by substituting Equations 6.14 and 6.26 into the expression for $[P]_\infty$, that for irreversible inhibition described by reaction mechanism 6.4, the constant value of [P] reached as t approaches ∞ ($[P]_\infty$) is described by Equation 6.27. Potency for irreversible inhibitors of this type is determined by the values of K_i, and k_5.

$$[P]_\infty = \frac{V_{max}[S]K_i}{k_5 K_m [I]}$$

Equation 6.27   Expression for the amount of product generated at infinite time for an irreversible inhibitor following a 2-step binding mechanism.

It can often be useful to preincubate enzyme and inhibitor, to gain a better estimate for k_6. It is also useful that following this preincubation, the slow rate means that substrate utilization is lower, allowing the time-course to be followed for longer.

Characterization, distinction between 1-step and 2-step mechanisms and assessment of irreversible inhibitor effectiveness is often accomplished by examining the observed rate constant for enzyme inactivation at different concentrations of inhibitor. For a 1-step mechanism, the secondary plot is linear, whereas for a 2-step mechanism, it follows a hyperbolic function as seen in slow-binding (Figures 6.17 and 6.18).

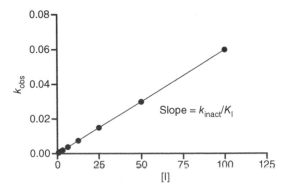

Figure 6.17   Relationship between k_{obs} and [I] in a 1-step binding inhibition mechanism.
A slow-binding inhibitor can bind to free enzyme and exert its inhibitory behavior in 1-step, when this occurs the enzyme slowly forms a complex with the inhibitor (EI) and the observed rate constant (k_{obs}) increases linearly as the inhibitor concentration increases.

Figure 6.18   Relationship between k_{obs} and [I] in a 2-step binding inhibition mechanism.
A slow-binding inhibitor can bind to free enzyme to form a complex (EI) quite rapidly, however, a slow isomerization of the enzyme into a higher affinity complex (EI*) can occur subsequently, so that inhibitor appears to bind slowly. Initial binding followed by an isomerization is a 2-step binding mechanism and the relationship between the rate (k_{obs}) and the inhibitor concentration ([I]) is hyperbolic.

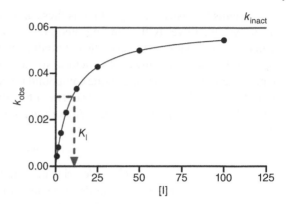

In the second, hyperbolic plot, the maximum observed rate of inactivation is termed k_{inact}, with the inhibitor concentration required for half of the maximal rate of inactivation, denoted as K_I. Note, this is not the same as K_i, from the Reaction scheme 6.4 and Equation 6.10, above, which describes the dissociation of the EI complex. In a similar manner to comparing the specificity of enzymes using k_{cat}/K_m, the efficacy of irreversible inhibitors can be compared using k_{inact}/K_I. This bimolecular rate constant accounts for the potency and the rate of inactivation. From inspection of the Reaction scheme 6.4, where $k_6 = 0$, the value of K_I is given by Equation 6.28, where $k_5 = k_{inact}$.

$$K_I = \frac{k_3 + k_5}{k_4}$$

Equation 6.28   Expression for K_I.
Here, K_I is the inhibitor concentration that yields the half maximal k_{obs} value from the plot of k_{obs} versus [I], as shown in Figure 6.18.

For the 1-step mechanism, the efficiency of the reaction may also be characterized by k_{inact}/K_I, where this parameter is now the slope of the linear relationship shown in Figure 6.17.

Caution should be applied in the characterization of 2-step mechanisms, because at concentrations of irreversible compound well below (K_I) a plot of k_{obs} versus [I] will be linear, passing through the origin, with slope $= k_{inact}/K_I$ similar to that for the 1-step mechanism. Hence, it is important to use a wide range of inhibitor concentrations during the analysis [10].

Key Example: Irreversible Inhibition by Tagrisso (Osimertinib)

Osimertinib is a covalent, third-generation epidermal growth factor receptor (EGFR) tyrosine kinase inhibitor used for the treatment of non-small cell lung cancer in patients with activating EGFR mutations. The drug achieves its mutant selectivity by inactivating the L858R single mutant and the L858R/T790M double mutant with 20-fold and 50-fold higher overall efficiencies (larger k_{inact}/K_I values), respectively, compared to the wild-type protein. It does this through increased initial affinity relative to competition by ATP and better positioning of the acrylamide warhead for reactivity [10].

6.10   Presence of Two Inhibitors

Sometimes it may be useful to know whether two different inhibitors affect the enzyme by interacting with the same site. Caution should be used when using kinetics- and ligand-binding experiments to answer this question as these approaches only measure the functional properties of the system. To verify the binding site(s) that two different compounds use ultimately requires structural information. However, ligand binding and kinetic approaches may be useful in providing useful information ahead of structural studies. For example, competition binding studies may demonstrate that the binding of one compound prevents the binding of a second. Enzyme kinetics may also be used to assess the effects of two inhibitors.

If in the case of inhibitor, I alone, $v_i/v_0 = i$ and for inhibitor X alone, $v_x/v_0 = x$, then in the presence of both I and X, $v_{ix}/v_0 = ix$. This principle allows the evaluation of complex rate equations as the expressions are simplified (Equation 6.29) when:

- $I = 0$ terms are lost, these terms describe inhibition by I.
- $X = 0$ terms are lost, these terms describe inhibition by X.
- I or $X = 0$, these terms account for association with both I and X, and these terms are lost when binding of I and X is mutually exclusive.

For example, consider the simple case from above, where

$$\frac{v_i}{v_0} = \frac{1}{\left(1 + \frac{[I]}{K_i'}\right)}$$

and

$$\frac{v_x}{v_0} = \frac{1}{\left(1 + \frac{[X]}{K_x'}\right)}$$

When binding is independent:

$$\frac{v_{ix}}{v_0} = \frac{1}{\left(1 + \frac{[I]}{K_i'} + \frac{[X]}{K_x'} + \frac{[I][X]}{K_i'K_x'}\right)}$$

When binding is mutually exclusive:

$$\frac{v_{ix}}{v_0} = \frac{1}{\left(1 + \frac{[I]}{K_i'} + \frac{[X]}{K_x'}\right)}$$

In the general case:

$$\frac{v_{ix}}{v_0} = \frac{1}{\left(1 + \frac{[I]}{K_i'} + \frac{[X]}{K_x'} + \frac{\alpha[I][X]}{K_i'K_x'}\right)}$$

Equation 6.29 **Expression for the rate of reaction, v_{ix}, for an enzyme reaction, compared to the control rate, v_0, in the presence of two inhibitors.**

Where α is an interaction factor. When α tends to zero binding is mutually exclusive. When $\alpha = 1$, binding is completely independent. When $\alpha < 1$, then the compounds interfere with each other but do not prevent the binding of each other. When $\alpha > 1$, the presence of one compound favors association with the other.

To analyze data for two inhibitors, it is useful to vary both [I] and [X] in a single experiment. However, distinguishing between independent and mutually exclusive compounds may be difficult, since the differences in rate between these two extremes is large only when both compounds

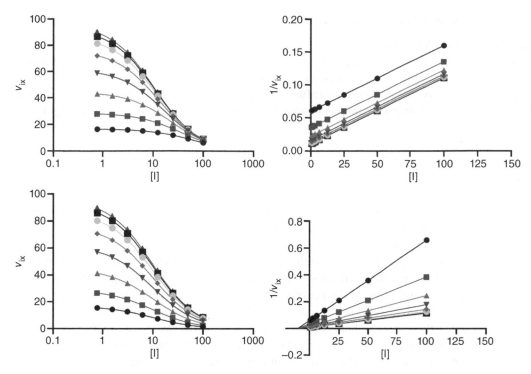

Figure 6.19 **Nonlinear and linear plots for mutually exclusive and independent binding between two inhibitors.**
Nonlinear plots (left panel) and Dixon plots (right panel) describing the behavior of enzyme rate in the presence of two inhibitors. The top panel shows the behavior for mutually exclusive inhibitors, with the bottom panel showing the behavior when the inhibitors are independent of each other.

are present at high concentrations, meaning that the observed rate is low and difficult to measure accurately.

Whilst the analysis should be undertaken by fitting an appropriate model using nonlinear regression analysis, the use of linear secondary plots can be useful to demonstrate mutually exclusive versus independent kinetics. Dixon plots (where $1/v_{ix}$ is plotted versus [I]) show non-intersecting lines for mutually exclusive kinetics and intersecting lines for independent kinetics (Figure 6.19).

6.11 Non-specific Inhibition

There are several ways molecules can act to inhibit enzymes in a manner that has often been described as non-specific inhibition. This term is perhaps misleading as often these molecules have specific mechanisms by which they inhibit yet are undesired and are not suitable for exploitation in drug discovery. We may be able to predict, avoid or identify this behavior to a lesser or greater extent depending upon the mechanism. Identifying the specific mechanism of "non-specific" inhibition may be useful for improving prediction, avoidance and identification of this behavior. If this type of behavior is not identified, there is a danger that optimization activities may be confounded due to following mixed structure–activity relationships (SAR).

6.11.1 Common Technology Hitters

6.11.1.1 UV-light Interference

Many assay technologies utilize light to measure the rate of the enzyme catalyzed reaction, see Table 6.3 for some common chromophores used in enzyme assays. Some test compounds may interfere with the detection system, to reduce the signal and so may appear to be inhibitors. For example, compounds may form minute precipitates in solution which scatter light. This light scattering effect may disrupt the intensity of light and reduce the apparent signal. These types of compounds will show less inhibition when the small particulates become solubilized, for example, at lower concentrations or in the presence of non-ionic detergents, for example, Tween 20, Triton X-100 or CHAPS. Alternatively, some test compounds may interfere with assay readouts by absorbing light at wavelengths that are used by the assay technology. For example, the electrons of colored compounds absorb energy from the incident UV light or emitted light and emit the remainder of energy as colored light. This effect increases at higher test compound concentration and can mimic inhibition in absorbance-based assays.

Test compounds may also interfere with fluorescence-based assays either by quenching the fluorescence of the detection system generated by the assay process, reducing the signal, or by showing autofluorescence, contributing to the generated signal. A list of common fluorophores used in enzyme assays is shown in Table 6.4. A compound may absorb light leading to a reduction in the intensity of either the excitation light or emitted light produced during the assay. The absorbance of either the light coming into the assay system or the light leaving the assay system will decrease the overall signal and may lead to false-positive or negative results, depending upon the assay configuration.

6.11.1.2 Detection System Interference

Some compounds may have chemical structures that lead to disruption of the expected formation of the interaction between components of a key capture system by similarity to the chemical structure of one of the reagents. An example is the interaction between biotin tags and streptavidin labels. Alternatively, some test compounds may inhibit enzymes, unrelated to the target, but which are used as coupling enzymes required to generate the assay signal. Coupled assays often have one or more enzymes present, which are required to convert the product for the enzyme under study to reagents that can be more easily measured. The presence of these additional enzymes means that compounds that do not inhibit the target enzyme, but which do inhibit the coupling enzymes, will be identified as active. Although the coupling enzymes are included at high concentrations to ensure that product generated by the target enzyme is immediately converted to detectable reagent,

Table 6.3 **Common chromophores employed in enzyme assays.**

Chromophore	Wavelength (nm)	Extinction coefficient ($M^{-1}\ cm^{-1}$)
NAD(P)H	340	6 220
4-Nitrophenol	400	21 000
4-Nitroaniline	405	9 620
DTNB	412	13 600
Malachite green	618	140 000

Table 6.4 **Common fluorophores employed in enzyme assays.**

Fluorophore	Excitation wavelength (nm)	Emission wavelength (nm)	Color of emitted light
Tryptophan	280	360	Outside visible range
AMC (7-Amido-4-methylcoumarin)	360	440	
EDANS (5-[(2-Aminoethyl)-amino]naphthalene-1-sulfonyl)	340	490	
Fluorescein isocyanate	490	520	
Rhodamine	497	521	
Cy3	554	565	
TAMRA	522	577	
Texas Red	594	613	
Cy5	648	665	
Alexa 647	650	670	

this highlights the need for testing active compounds in the detection system alone, without the target enzyme.

6.11.2 Chemical Reactivity

Many compounds display reactivity and although these compounds are often removed from screening collections, there may still be those that react non-specifically with some enzyme targets. There are several functional groups that are considered as potentially problematic and should be avoided in chemical structures taken forward for optimization. These include alkyl and acyl halides, aziridines, anhydrides, alkyl sulfonates, isocyanates, peroxides, triflates, carbodiimides, disulfides, thiols, epoxides, aziridines, acyl and sulfonyl cyanides, aldehydes, reactive Michael acceptors, ketenes, carbamic and boronic acids, fluoro-pyridines, nitro-aromatics and heteroaromatics, beta-lactones, beta-lactams, activated esters and imines, hydrazines, cyclohexadienes, and hydroxy aryl-anilines.

6.11.3 Aggregators

Many compounds contained within screening libraries are known to form aggregates at concentrations often used in screening assays, see Figure 6.20. These aggregates subsequently absorb onto the enzyme surface or sequester enzymes within them. Inhibition by this mechanism is characterized by some or more of the following characteristics: reversible, but time-dependent; steep concentration–response curves; temperature/denaturant independent; enzyme concentration, ionic strength and detergent/BSA dependent; non-competitive kinetics [11]. Detecting these types of compounds involves understanding this behavior by undertaking experiments to observe the effect of compound pre-incubation on the degree of inhibition observed, ensuring that steep slopes are not masked by incorrect fitting, observing the effect of detergent on the inhibitor behavior and observing the effects of the compound on other enzyme systems under similar conditions of concentration and buffer composition. Since few experiments are definitive alone, a combination of observations in the target assay and subsequent tests are recommended for understanding this type of behavior: mM IC_{50}; Steep concentration–response; IC_{50} increases with addition of detergent; time-dependent IC_{50}; IC_{50} enzyme concentration dependency, which is not tight binding; activity comparable against unrelated enzyme (chymotrypsin, for targets other than proteases). There are also preventative measures that could be considered. For example, using predictive algorithms based on calculated Log P values, including the presence of a nitrogen atom bonded to three heavy atoms, the presence of carboxyl groups and the degree of conjugation [12]. Assays can be optimized to reduce the likelihood of detecting activity due to aggregation. Steps that can be taken include adding detergent to the assay or increasing the ionic strength. Understanding compounds before including in the assay, for example, by annotating compounds previously having shown aggregation. Alternatively, a screen for aggregation could be undertaken, or compounds could be tested for their ability to inhibit enzymes unrelated to the target, for example, chymotrypsin or AmpC β-lactamase [11].

Key Example: Enzyme Inhibition by Aggregation

One of the main sources of false positives in primary screening is the formation of small molecule aggregates, which inhibit enzymes non-specifically at the micromolar concentrations typically used in the screen. The small molecule rottlerin has been studied as a model aggregator and its inhibitory activity versus AmpC β-lactamase has been ascribed to partial unfolding of the protein following binding to the aggregate particle [21].

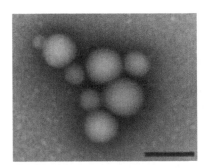

Figure 6.20 Transmission EM of stabilized fulvestrant colloids. The figure shows an electron micrograph of colloids formed by the drug fulvestrant. Although this is a marketed drug, this demonstrates that compounds can form aggregates and secondly that the biological activity may be separate from the formation of aggregates. It is therefore important to understand the relevant concentrations at which inhibition by a specific mechanism and that caused by unwanted mechanisms, such as aggregation, occur.

6.11.4 Redox Reactivity

Some compounds may react with the reducing agent present in the assay (e.g. DTT) to form damaging free radicals (e.g. superoxide radical anions or hydroxyl radicals are particularly damaging). For example, a compound Q may react with a reducing agent thiol (R–SH), following the following chain reaction to generate the highly damaging superoxide radical.

$$Q + R\text{–}SH \rightarrow Q^{-\cdot} + H^+ + R\text{–}S^\cdot$$

$$R\text{–}S^\cdot + R\text{–}SH \rightarrow R\text{–}S\text{–}S\text{–}R^{-\cdot} + H^+$$

$$R\text{–}S\text{–}S\text{–}R^{-\cdot} + Q \rightarrow Q^{-\cdot} + R\text{–}S\text{–}S\text{–}R$$

$$R\text{–}S^\cdot + Q \rightarrow R\text{–}SQ^\cdot$$

$$R\text{–}SQ^\cdot + R\text{–}SH \rightarrow Q^{-\cdot} + R\text{–}S\text{–}S\text{–}R + H^+$$

$$Q^{-\cdot} + O_2 \rightarrow Q + O_2^{-\cdot}$$

Or the presence of reduced transition metals (e.g. Fe^{2+}, Cu^{2+}, Cr^{3+}), which may be present in trace amounts, may promote radical formation:

$$Fe^{2+} + O_2 \rightarrow Fe^{3+} + O_2^{-\cdot}$$

$$2O_2^{-\cdot} + 2H^+ \rightarrow H_2O_2 + O_2$$

$$Fe^{2+} + H_2O_2 \rightarrow Fe^{3+} + OH^\cdot + OH^-$$

The reaction of Fe^{2+} with hydrogen peroxide to form the hydroxyl radical and the OH^- is called the Fenton reaction.

Free radicals are species that contain an unpaired electron, and so are extremely chemically reactive. This is due to the need for the radical to pair its single electron. Thus, a free radical must abstract an electron from an adjacent molecule. This causes the formation of another free radical and a chain reaction begins. The superoxide and hydroxyl radicals generated in the reactions shown above can be highly damaging to proteins, causing oxidation of side chains, cross-linking, and oxidation of the protein backbone, leading to inactivation. Sulfur-containing amino acids (cysteine and methionine) are extremely susceptible to oxidation, with cysteines converted to disulfides.

6.11.4.1 Oxidation of Aromatic Amino Acid Residues
The oxidation products of the aromatic amino acids are illustrated in Table 6.5.

Typically, compounds may show time-dependent inhibition in the presence of DTT or TCEP, but often no effect when glutathione or cysteine is used.

Compounds showing this behavior can be identified by following the reaction spectrophotometrically when compound and reducing agents, such as DTT, are incubated.

6.11.4.2 Protein Unfolding
Some compounds may be active by binding to, and stabilizing, the unfolded state (Reaction scheme 6.5).

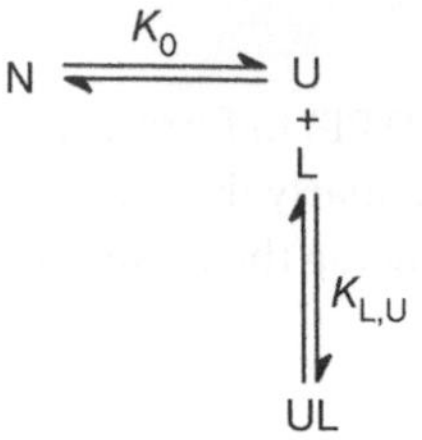

Reaction scheme 6.5 Reaction scheme showing ligand binding to and stabilizing unfolded protein.

Table 6.5 Oxidation product5s of aromatic amino acids.

Sidechain	Oxidative products			

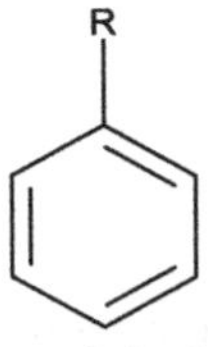

Phenylalanine

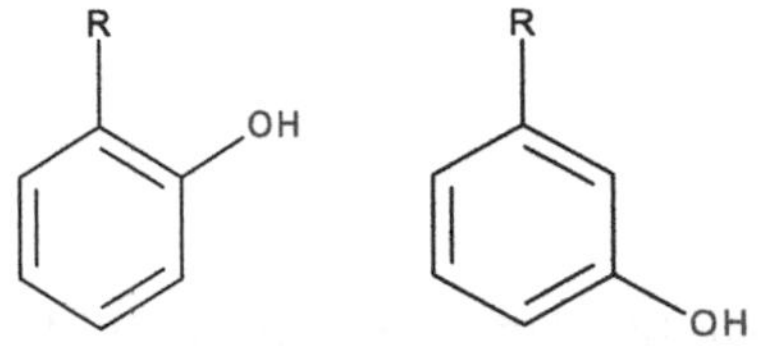

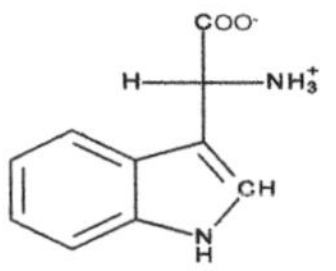

Tryptophan

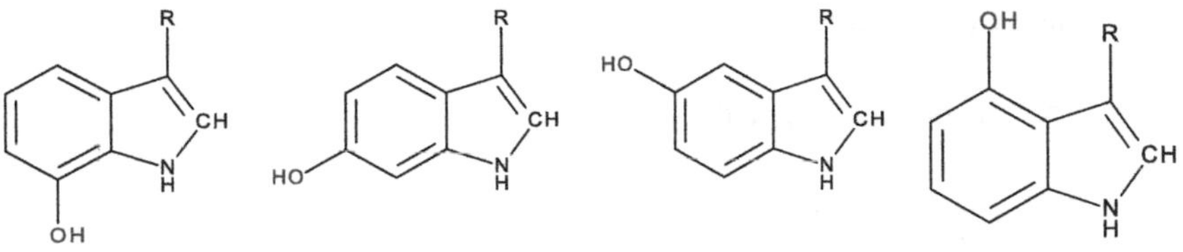

Tyrosine

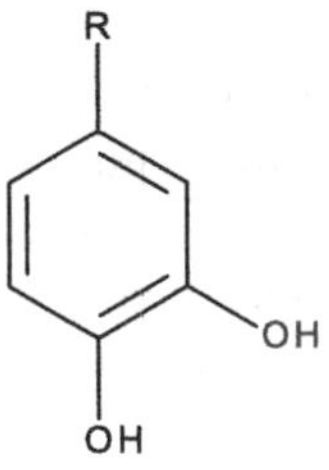

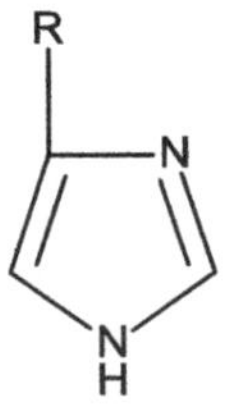

Histidine

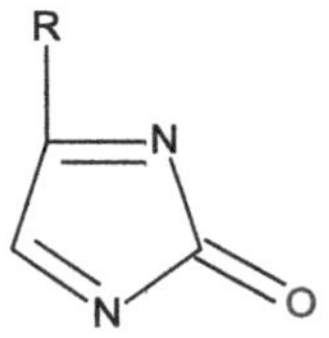

Figure 6.21 Energy profile diagram for protein unfolding.
Free energy diagram showing ligand binding to the unfolded
state of a protein, stabilizing the unfolded form. A lower energy,
ΔG_{obs}, is therefore required to populate the unfolded state,
compared to the absence of ligand.

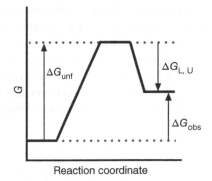

Reaction coordinate

Reaction scheme 6.5 above shows how ligands may bind to the unfolded state (U) of the protein and perturb the equilibrium between native (N) and unfolded states. The free energy diagram for this behavior is shown below (Figure 6.21).

These compounds may be detected as a decrease in stability of the protein, utilizing methods such as DSC or DSF.

6.11.5 Denaturation

Some compounds may denature proteins not by binding directly or by forming aggregates themselves, but by indirect methods, such as disruption of electrostatic interactions, disruption of hydrophobic interactions, disruption of water–water interactions and oxidation–reduction. These effects may increase the density (large numbers of water molecules containing many weak bent and/or broken hydrogen bonds) of the first hydration shell, allowing protein self-association and inactivation.

Chaotropes tend to have large singly charged ions, for example, SCN^-, $H_2PO_4^-$, HSO_4^-, HCO_3^-, I^-, Cl^-, NO_3^-, NH_4^+, K^+, guanidinium and tetramethylammonium. Perhaps, the most unfavorable situation is having a chaotropic anion and a kosmotropic cation (note chaotropes unfold proteins, destabilize hydrophobic aggregates, and increase the solubility of hydrophobic molecules, whereas kosmotropes stabilize proteins and hydrophobic aggregates in solution and reduce the solubility of hydrophobic molecules), which leads to salting out conditions. This could lead to decreased stability of already unstable enzyme domains. Detecting chaotropic or denaturation effects may be accomplished using approaches for compounds unfolding the target protein, where decreased stability of the folded form will tend to give reduced T_m shifts in techniques, such as DSC and DSF.

6.11.6 Metal Ion Contamination

It is well known that organic impurities present in a compound sample can cause false-positive signals in hit identification approaches, when pure compound shows no activity. However, the presence of organic contamination does not fully explain all false-positive results when an impurity is the cause. Inorganic impurities also can lead to positive signals for many targets and detection systems. Metal ion impurities often originate during compound synthesis, as transition metal catalysts and metal-containing precursors are frequently used during synthesis methods. Even following purification procedures to remove contamination, small traces of metal ions may remain in some compound samples. Although the analysis of compound identity by NMR and/or mass spectrometry allows the detection of organic breakdown products, they do not reveal the presence of potentially problematic metal ions. Furthermore, these metal-containing samples may also demonstrate positive results in some orthogonal assays designed to evaluate primary screening output.

Metal ions may bind in the active site and disrupt catalytic function. Additionally, some metals, for example, silver, mercury, copper, and lead may react with thiol groups in the side groups of cysteine residues. This may lead to conformational changes resulting in incorrect folding which is not compatible with correct enzyme function.

Whilst the use of chelators to bind metals is well established, for example, using ethylenedinitrilotetraacetic acid (EDTA) to chelate divalent metal ions, some assays cannot tolerate the inclusion of EDTA. Additionally, EDTA does not enable the removal of all potential interfering metal ions. Typically, approaches to detect metal ion contamination are introduced during the active-to-hit stage, where the numbers of compounds to test have been reduced [13].

References

1 Reed MC, Lieb A, Nijhout HF. The biological significance of substrate inhibition: a mechanism with diverse functions. *BioEssays: News and Reviews in Molecular, Cellular and Developmental Biology*. 2010;32(5):422–9.

2 Kosow DP, Rose IA. Product Inhibition of the Hexokinases. *Journal of Biological Chemistry*. 1970;245(1):198–204.

3 Christopoulos A. Assessing the distribution of parameters in models of ligand-receptor interaction: to log or not to log. *Trends in Pharmacological Sciences*. 1998;19(9):351–7.

4 Naski MC, Fenton JW, Maraganore JM, Olson ST, Shafer JA. The COOH-terminal domain of hirudin. An exosite-directed competitive inhibitor of the action of alpha-thrombin on fibrinogen. *Journal of Biological Chemistry*. 1990;265(23):13484–9.

5 Zerner B, Bender ML. The kinetic consequences of the acyl-enzyme mechanism for the reactions of specific substrates with chymotrypsin. *Journal of the American Chemical Society*. 1964;86(18):3669–74.

6 Wong CH. Use of Hydrolytic Enzymes: Amidases, Proteases, Esterases, Lipases, Nitrilases, Phosphatases, Epoxide Hydrolases. In: *Enzymes in Synthetic Organic Chemistry*: Elsevier Science Ltd.; 1994:41–130.

7 Morrison JF, Walsh CT. The Behavior and Significance of Slow-binding Enzyme Inhibitors. In: *Advances in Enzymology – and Related Areas of Molecular Biology*: Wiley; 1988. p. 201–301.

8 Ward WHJ, Holdgate GA, Rowsell S, McLean EG, Pauptit RA, Clayton E, et al. Kinetic and structural characteristics of the inhibition of enoyl (acyl carrier protein) reductase by triclosan. *Biochemistry*. 1999;38(38):12514–25.

9 Luckner SR, Liu N, Am Ende CW, Tonge PJ, Kisker C. A slow, tight binding inhibitor of InhA, the enoyl-acyl carrier protein reductase from *Mycobacterium tuberculosis*. *The Journal of Biological Chemistry*. 2010;285(19):14330–7.

10 Zhai X, Ward RA, Doig P, Argyrou A. Insight into the therapeutic selectivity of the irreversible EGFR tyrosine kinase inhibitor osimertinib through enzyme kinetic studies. *Biochemistry*. 2020;59(14):1428–41.

11 Coan KED, Maltby DA, Burlingame AL, Shoichet BK. Promiscuous aggregate-based inhibitors promote enzyme unfolding. *Journal of Medicinal Chemistry*. 2009;52(7):2067–75.

12 Seidler J, McGovern SL, Doman TN, Shoichet BK. Identification and prediction of promiscuous aggregating inhibitors among known drugs. *Journal of Medicinal Chemistry*. 2003;46(21): 4477–86.

13 Molyneux C, Sinclair I, Lightfoot HL, Walsh J, Holdgate GA, Moore R. High-throughput detection of metal contamination in HTS outputs. *SLAS Discovery*. 2022;27(5):323–9.

14 Kettner CA, Shenvi AB. Inhibition of the serine proteases leukocyte elastase, pancreatic elastase, cathepsin G, and chymotrypsin by peptide boronic acids. *The Journal of Biological Chemistry*. 1984;259(24):15106–14.

15 Shapiro R, Riordan JF. Inhibition of angiotensin converting enzyme: mechanism and substrate dependence. *Biochemistry* 1984;23(22):5225–33.

16 Baici A, Camus A, Marsich N. Interaction of the human leukocyte proteinases elastase and cathepsin G with gold, silver and copper compounds. *Biochemical Pharmacology*. 1984;33(12):1859–65.

17 Nakamura CE, Abeles RH. Mode of interaction of beta-hydroxy-beta-methylglutaryl coenzyme A reductase with strong binding inhibitors: compactin and related compounds. *Biochemistry*. 1985;24(6):1364–76.

18 Stone SR, Morrison JF. Mechanism of inhibition of dihydrofolate reductases from bacterial and vertebrate sources by various classes of folate analogues. *Biochimica et Biophysica Acta (BBA) – Protein Structure and Molecular Enzymology*. 1986;869(3):275–85.

19 Ward WH, Holdgate GA, Freeman S, McTaggart F, Girdwood PA, Davidson RG, et al. Inhibition of squalene synthase in vitro by 3-(biphenyl-4-yl)-quinuclidine. *Biochemical Pharmacology*. 1996;51(11):1489–501.

20 Rich DH, Sun ETO. Mechanism of inhibition of pepsin by pepstatin: effect of inhibitor structure on dissociation constan and time-dependent inhibition. *Biochemical Pharmacology*. 1980;29(16):2205–12.

21 Moorman AR, Abeles RH. A new class of serine protease inactivators based on isatoic anhydride. *Journal of the American Chemical Society*. 1982;104(24):6785–6.

22 Gupton BF, Carroll DL, Tuhy PM, Kam CM, Powers JC. Reaction of azapeptides with chymotrypsin-like enzymes. New inhibitors and active site titrants for chymotrypsin A alpha, subtilisin BPN', subtilisin Carlsberg, and human leukocyte cathepsin G. *Journal of Biological Chemistry*. 1984;259(7):4279–87.

23 Furman PA, St. Clair MH, Spector T. Acyclovir triphosphate is a suicide inactivator of the herpes simplex virus DNA polymerase. *Journal of Biological Chemistry*. 1984;259(15):9575–9.

7

Enzyme Activation and Its Comparison with Inhibition

7.1 Introduction

Activators are molecules that are capable of binding to an enzyme and increasing its activity. The presence of an activator in an enzyme assay increases the rate of the enzyme-catalyzed reaction by binding to the enzyme and accelerating the conversion of substrate to product. Activators may also function by binding to the substrate, where the substrate-activator complex is the true substrate, but this behavior will not be discussed here. Activators may be essential for activity or may raise the activity from a basal level in a concentration-dependent manner, where higher concentrations of activator result in higher levels of activity until a maximum level of activation is reached, and the enzyme is fully saturated by activator and substrate. These different types of activation behavior will be discussed below.

7.2 Mechanisms for Enzyme Activation

There are two distinct types of enzyme activation; the first where the enzyme requires the presence of an activating molecule for catalytic activity and the second where there is a basal activity that can be enhanced by the presence of activator. These different types of activation are classified as essential and non-essential activation, respectively.

7.2.1 Essential Activation

Essential activation describes the situation where a small molecule, or often a cation, is absolutely required, alongside the substrate, for catalysis to occur. The enzyme remains in an inactive state and cannot turn over substrate until it is activated by the binding of the small molecule activator. Binding of the activator often results in a conformational change, which reorientates the catalytic machinery and switches the enzyme from the inactive to the active state. This normally involves a large entropic penalty during rearrangement of the site [1, 2]. Similar to linear inhibition, there

Laboratory Guide to Enzymology, First Edition. Geoffrey A. Holdgate, Antonia Turberville, and Alice Lanne.
© 2024 John Wiley & Sons, Inc. Published 2024 by John Wiley & Sons, Inc.

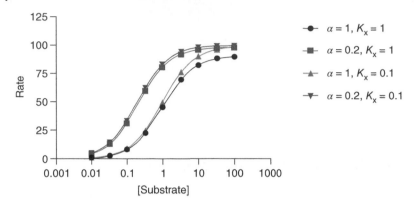

Figure 7.1 Substrate dependence plot for different essential activator parameters.
The plot shows the effect of activators with different parameter values, where $V_{max} = 100$, $K_s = 1$, $[A] = 10$, and the parameter values for the four hypothetical essential activators are as shown in the legend.

are two prominent mechanisms for essential activation, which are characterized by the reciprocal allosteric coupling value, α. Essential activators may bind to free enzyme only (where $\alpha K_x = \infty$, where K_x is the equilibrium dissociation constant for activator binding to free enzyme) or have mixed behavior, binding before substrate to free E or after substrate to the ES complex (Reaction scheme 7.1). There are essentially three subsets of mixed behavior: preferential binding to free enzyme ($1 < \alpha < \infty$); preferential binding to the ES complex ($0 < \alpha < 1$) and equal affinity for binding to either enzyme form ($\alpha = 1$). Usually, small molecule essential activators are not a focus for drug discovery, as restoration or improvement in the rate of enzyme activity is often required rather than the *ab initio* generation of substrate turnover.

$$
\begin{array}{ccccc}
\text{E} & + & \text{S} & \overset{K_s}{\rightleftharpoons} & \text{ES} \\
+ & & & & + \\
\text{A} & & & & \text{A} \\
\Big\updownarrow K_x & & & & \Big\updownarrow \alpha K_x \\
\text{EA} & + & \text{S} & \overset{\alpha K_s}{\rightleftharpoons} & \text{EAS} \xrightarrow{k_{cat}} \text{EA} + \text{P}
\end{array}
$$

Reaction scheme 7.1 General scheme for essential activation.
In essential activation, only the enzyme-activator-substrate complex turns over to yield a product. K_s is the substrate equilibrium dissociation constant, K_x is the equilibrium dissociation constant for activator. The apparent K_m is αK_s.

The rate equation for essential activation is given by Equation 7.1, below.

$$
v = \frac{V_{max}\,[S]}{\alpha K_s \left(1 + \frac{K_x}{[A]}\right) + [S]\left(1 + \frac{\alpha K_x}{[A]}\right)}
$$

Equation 7.1 Expression for the rate equation for an essential activator.

It can be seen from Equation 7.1 that at very high activator concentrations, where $[A] \gg K_x$, the rate equation reduces to the Michaelis–Menten equation with $K_m = \alpha K_s$.

The effect of essential activators with varying parameters is shown in Figure 7.1 below.

7.2.1.1 Essential Cationic Activation

As mentioned above, often essential activators are cations. In this case, a monovalent or divalent metal ion is required for catalysis and the enzyme cannot turn over substrate even at high concentrations, in the absence of the cation [1]. Essential cationic activation is often described as type I cationic activation and may be further divided into two subsets, type Ia and Ib [1, 3, 4]. Type Ia usually involves the metal ion anchoring the substrate into the active site. The metal ion binds to free enzyme, and coordinates and rearranges the active site for catalysis by enhancing substrate binding. Enzymes, such as diol and glycerol dehydratases, are activated by monovalent cations through a type Ia mechanism [5–7].

Type Ib activation occurs when the metal binding site is distinct from the substrate recognition site. In this case, the metal ion may bind to free enzyme or the ES complex. This represents a "non-competitive" activation mechanism. Pyruvate kinase is activated by potassium ions (K^+) and k_{cat} is zero when K^+ is not present. K^+ can bind to different forms of pyruvate kinase; active (closed) or inactive (open). This demonstrates that the activation mechanism is type Ib [8–11].

7.2.2 Non-essential Activation

In drug discovery, where an increase in enzyme activity is desired, non-essential activators are often sought. Non-essential activation is observed when the enzyme can turn over the substrate in the absence of the activator, although the rate is lower than with activator present. As with essential activators, a non-essential activator may bind either before substrate, to the free enzyme, or after the substrate, to the ES complex (Reaction scheme 7.2). Activator binding leads to either or both improved substrate binding and/or acceleration of the chemical step, an example of this activation is seen with glucokinase [12–20]. The mechanisms of non-essential activation are discussed further in Chapter 8.

$$
\begin{array}{ccccccc}
E & + & S & \xrightleftharpoons{K_m} & ES & \xrightarrow{k_{cat}} & E + P \\
+ & & & & + & & \\
A & & & & A & & \\
\big\Updownarrow K_x & & & & \big\Updownarrow \alpha K_x & & \\
EA & + & S & \xrightleftharpoons{\alpha K_m} & EAS & \xrightarrow{\beta k_{cat}} & EA + P
\end{array}
$$

Reaction scheme 7.2 General scheme for non-essential activation.
In non-essential activation, the enzyme-substrate complex turns over to product in the absence of an activator. For activation to occur, in general, $\alpha < 1$ and/or $\beta > 1$. Mixed activation may occur with a range of α and β values, which is discussed more in Chapter 8.

The rate equation for non-essential activation is given by Equation 7.2, below.

$$
v = \frac{V_{max}[S]\left(1 + \frac{\beta[A]}{\alpha K_x}\right)}{K_m\left(1 + \frac{[A]}{K_x}\right) + [S]\left(1 + \frac{[A]}{\alpha K_x}\right)}
$$

Equation 7.2 Expression for the rate equation for a non-essential activator.

The substrate dependence curves for the presence and absence of non-essential activators with a range of different parameters are shown in Figure 7.2.

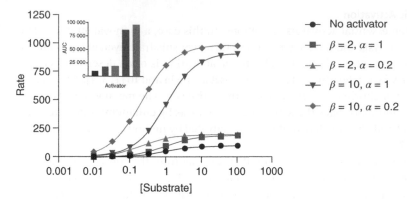

Figure 7.2 **Substrate dependence plots in the presence and absence of activator.**
The black line shows the substrate dependence in the absence of non-essential activator, with parameter values $V_{max} = 100$, $K_m = 1$. The colored lines show the substrate dependence in the presence of activator, $[A] = 10$, with the values of α and β as shown in the legend. Inset shows the area under the curve (AUC) for each curve as described in the legend.

The degree of activation is affected by changes in any of the parameters α, β, and K_x, where lower values of α and K_x compared to a reference compound lead to enhanced activation. β values which are increased compared to a reference compound also lead to enhanced activation. The effect of different values of the parameters α, β, and K_x on the activator concentration response is shown in Figure 7.3.

7.2.2.1 Non-essential Cationic Activation

> **Key Example: Non-essential Small Molecule Activators**
>
> Glucokinase (GK) is an attractive therapeutic target for modulating blood glucose homeostasis in type II diabetes. Mutation in GK frequently can cause loss or gain of function where loss of function mutations result in a reduction in substrate affinity (glucose and ATP) and reduced sensitivity of β cells to glucose and gain of function corresponds to increased catalytic activity and hyperinsulinemic hypoglycaemia [12, 13]. A compound, identified from a compound screen was found to enhance the basal activity of GK with a 4-fold decrease in K_m' and 1.5-fold increase in V_{max}. No effect on the K_m' for ATP (the second substrate) was observed [21]. This modifier shows both V- and K-type activation and is characteristic of "uncompetitive" non-essential activation.
>
> The discovery of this activator led to the development of a series of different activators [14–17]. These compounds typically bind to an allosteric site on the enzyme and stabilize the active conformation. This allosteric site is uncovered when GK binds glucose through a conformational change, the activator occupies the allosteric site preventing the enzyme from returning to its original conformation [18–20].

Non-essential cationic activation is also known as type II cationic activation and occurs when the presence of a cation increases activity, but the enzyme can still turn over substrate in the absence of the cation [1]. This is similar to type Ib cationic activation. However, for type II cationic activators, the enzyme will have activity in the absence of the specified cation. For example,

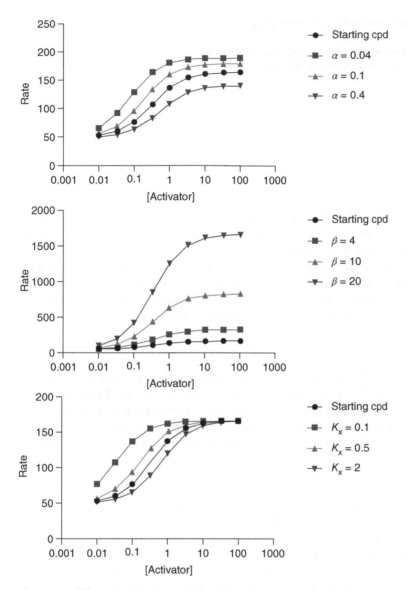

Figure 7.3 Effect of changes in α, β, and K_x on the degree of activation.
Starting compound conditions (starting cpd) are $\alpha = 0.2$, $\beta = 2$, $[S] = K_s = K_x = 1$, $V_{max} = 100$ (black circles).
The effects of changes in α, β, and K_x are shown in each panel with the values of the parameters indicated in the legends, respectively.

activation of branched-chain α-ketoacid dehydrogenase (BCKD) by K^+, involves binding to two separate allosteric sites, which stabilize the enzyme and allow maximal activity [21, 22].

7.2.3 A Comment on Nomenclature: K-type and V-type Classification

A simple way to classify activator (and inhibitor) modifiers that have been used in the literature is by their ability to affect the K_m or V_{max} of the enzymatic catalyzed reaction. Compounds decreasing K_m are classified as K-type activators, whilst compounds increasing V_{max} of the enzymatic

Table 7.1 Classical characterization of K- and V-type regulators.

Type of regulator	Predominant effect on the corresponding macroscopic kinetic parameter
K-type activator	$\downarrow K_m$
V-type activator	$\uparrow V_{max}$
K-type inhibitor	$\uparrow K_m$
V-type inhibitor	$\downarrow V_{max}$
Neutral	No effect

reaction in a concentration-dependent manner are classified as V-type activators. Conversely, if a compound increases K_m and decreases V_{max} of the reaction, then this represents as K-type or V-type inhibition, respectively (Table 7.1). However, caution should be applied as there are cases where a modifier compound may demonstrate K-type activation and V-type inhibition or K-type inhibition and V-type activation and so does not distinguish whether the modifier is an activator, inhibitor or has dual behavior, depending upon the conditions of substrate concentration, and so is usually accompanied by supporting information to help discern modifier behavior [2, 23].

7.2.4 De-inhibition

Activation of enzymes also may be achieved by relieving inhibition brought about by regulatory domains. This usually occurs due to activator binding causing a conformational change in the regulated enzyme to relieve the observed inhibition. For example, kinases, such as adenosine monophosphate-activated protein kinase (AMPK), frequently contain such regulatory or autoinhibition domains (AID) that control access to the active site on the catalytic domain [20, 24–27].

Key Example: De-inhibition

AMPK is regulated by protein-protein interactions; the regulatory domain ($\beta\gamma$ complex) interacts with the catalytic domain (α subunit), holding the kinase in an inactive conformation. AMP binds to the regulatory domain, it disrupts the protein–protein interactions between the α subunit and $\beta\gamma$ complex allowing AMPK to become active. This behavior can be treated as de-inhibition. Modulation of AMPK is seen as an attractive therapeutic target and could treat metabolic syndromes and type II diabetes [20, 24, 25]. A-592107, a small molecule discovered from a high-throughput screen [26], binds to the regulatory domain and disrupts the interaction between the α and $\beta\gamma$ domain in a similar way to AMP, but at an alternative site [27]. This compound activates AMPK with a slightly improved EC_{50} value of 38 μM, compared to AMP (56 μM), but no improvement in V_{max} was observed. Whilst there is no overall increase in V_{max}, the presence of the activating compound triggers increased levels of enzyme activation at lower non-saturating concentrations of substrate.

7.3 Challenges for Identifying Non-essential Enzyme Activators

Drug discovery approaches may seek to identify non-essential activators that restore or enhance the basal level of enzyme activity to modulate a disease state. Putative activators may be identified using many of the existing traditional methods applied to inhibitors, but there are additional factors for activators that should be considered [28].

7.3.1 Enzymes have Evolved to be Active

Firstly, enzymes have evolved specifically to increase the rate of chemical reactions in biological systems. This makes the probability of identifying small molecules that can further increase catalytic rate potentially lower than for inhibitors (it is presumably more likely that molecules will interfere with catalysis rather than improve upon it).

7.3.2 Lack of Tool Compounds

Secondly, tool compounds often are essential for developing and optimizing enzyme assays. Having a reagent that can reproduce the desired outcome expected for a suitable hit is useful in defining the response of the assay and the potential activity of identified hits. A lack of tool compounds is likely to be greater for activators than for inhibitors, at least since substitutes for eliminating activity are more readily devised than for increasing activity. For inhibitors, control blank rates may be obtained in various ways: omitting enzyme, substrate, or another essential component, using denatured (boiled) enzyme, adding stop solution (e.g. low or high pH) before the reaction is started, using chelators to remove essential metal ions or specific or non-specific inhibitors. This allows estimation of the anticipated decrease in rate when the enzyme is fully inhibited. However, determining the estimated degree of enzyme activation is less easy. It can be seen from Figures 7.1 and 7.2 that the level of activation conveyed by a compound is dependent upon the degree of saturation as well as the magnitude of the effect on k_{cat}, which contrasts with the effect of simple linear inhibitors. Usually, the bottom of the plot of rate versus inhibitor concentration is well defined, whereas the top of the curve for an activator is effectively unknown and may be different for each activating ligand. There are steps that can be taken, for example, activation may be mimicked by increasing the functional enzyme concentration or by increasing the substrate concentration, but the magnitude of the effect may bear no relationship to that observed for different activators.

7.3.3 Maintaining Steady State

Another aspect that should be considered is that an enzyme activator, by definition, will increase the rate of the catalyzed reaction compared to the rate in the absence of activator. This will change the time course of the reaction and may reduce the length of the steady-state period. Activators lead to the substrate concentration changing more quickly and so additional effort may be required during assay development to ensure that the expected degree of activation allows for remaining in the steady-state and for initial rate measurements to be reliable for detecting activators. This may mean that assays need to run for shorter times. On the other hand, activators showing slow-binding kinetics may require longer incubation times to allow full equilibration, especially as the relative magnitude of the change in rate is likely to be much lower. For example, it may not be surprising for an inhibitor to show 90% inhibition (10-fold reduction in rate), whereas similar magnitude increases in enzyme activity are likely to be rare. Often compounds providing around 2-fold activation are considered respectable hits [29].

7.3.4 Mechanistic Considerations

The mechanism of activation may also be important for detecting activators. Like inhibitors, activators may interact with the enzyme before or after the substrate has bound, although, for activators,

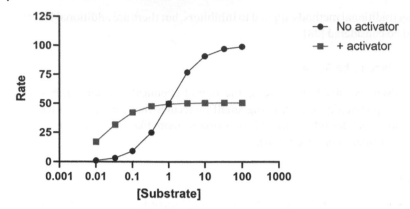

Figure 7.4 **Effect of substrate concentration in identifying mixed type activators – inhibition at high [S].** In the example above, the choice of substrate concentration has a marked effect on the ability to detect activation. At substrate concentrations below K_m, activation may be observed, with the greatest activation observed at $[S] = K_m/10$. No effect is observed at $[S] = K_m$ and inhibition is detected at concentrations above K_m. Parameter values were $\alpha = 0.02$, $\beta = 0.5$, $[A] = 10$, $K_x = K_m = 1$, $V_{max} = 100$.

the ternary complex between the enzyme, activator, and substrate must form for improved substrate turnover to occur. In general, the α and β terms from Reaction scheme 7.2, above, will often be below 1 and above 1, respectively. However, mixed-type activation is possible where the values of both α and β may both be above 1 or below 1. For the situation where both α and β are less than 1, it is possible for a compound to activate at low substrate concentrations but to show inhibition at higher concentrations. This can make the choice of substrate affecting the ability to detect activators with the desired mode of action [30], even more important than for inhibitors (see Figure 7.4).

Where both α and β are greater than 1, then the activator may demonstrate inhibition at low substrate concentrations, showing activation only at higher concentrations of substrate, Figure 7.5.

7.3.5 Assay Design and Variability

Identifying small molecule enzyme activators generally requires overcoming more challenges than the corresponding approaches for inhibitors. Many biochemical assay technologies that are

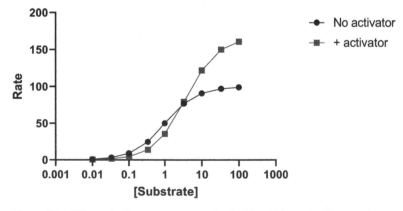

Figure 7.5 **Effect of substrate concentration in identifying mixed type activators – inhibition at low [S].** In this example, the compound shows inhibition at lower substrate concentrations, only demonstrating activation at concentrations of substrate several-fold above K_m. Parameter values were $\alpha = 5$, $\beta = 2$, $[A] = 10$, $K_x = K_m = 1$, $V_{max} = 100$.

employed for high throughput screens to detect inhibition also are amenable for identifying activators. However, assay quality parameters such as Signal Window, and Z' [31] may be estimated with lower certainty, due to higher backgrounds and more variable top signal. The assay window may actually be difficult to quantify for activator assays compared to inhibition assays due to the uncertainty over the degree of activation and the maximal signal achieved.

In cases where the standard deviation increases with a raw signal increase, variability around the minimum signal controls would be larger for activation assays since in this case, the minimum signal represents the basal level of activity rather than fully inhibited signal, and so is expected to be larger. Conversely, in inhibitor assays, the two controls correspond to 0% and 100% inhibition, where 0% inhibition is the uninhibited rate and 100% inhibition represents fully inhibited enzyme, often accessed by using a tool compound at high concentration. Access to these controls provides a defined signal window with easily quantifiable and often lower variability.

In contrast, for activator assays, the minimum control, as mentioned above, corresponds to the non-activated basal rate. The high signal control requires a tool compound, if available, at a concentration where saturation is reached, but it may be possible to reach higher signal levels if more effective activators are identified during the screen. Assuming the variance increases proportionally with signal increase (i.e. constant coefficient of variation, see Figure 9.3), these higher signal controls will have a higher standard deviation compared to the lower signal controls experienced in inhibition assays. Hence, there may be a tendency for the assay statistics to be poorer for activation compared to inhibition assays. This results in a lower probability for identifying compounds outside of the screening noise for activation assays.

Where tool compounds are not available, it may be necessary to use other controls to define the signal window. This is often simpler for inhibition assays, where the fully inhibited rate may be mimicked by various approaches as described above. For activation assays, in the absence of a tool-activating compound, the only option to mimic the activated rate, without affecting the chance of identifying activators with different mechanisms is to increase the enzyme concentration. Assessing the extent to which this may be a relevant approach is also difficult, as it is not easy to predict what degree of activation may be expected. The choice of an appropriate increase in enzyme concentration to provide a reasonable signal window estimate is therefore unclear. Even if a tool compound can be identified, there remains the distinct possibility for active compounds to have higher degrees of activation and therefore to display >100% activation with respect to the tool compound control. This uncertainty in assay window and the maximal activation potential, as well as the probability of higher variability, makes assay characterization and subsequent data analysis more difficult for activation assays.

Once active compounds are identified, concentration-response analysis also presents further issues for activator compounds. The most commonly used 4-parameter logistic fit model requires well-defined curves to reliably estimate the top, bottom, midpoint, and slope at the mid-point. As mentioned above, the top of an activator curve is often poorly defined by the data and more difficult to estimate.

This, therefore, results in more uncertain estimates for activator potency, requiring alternative methods to assess the effectiveness of active compounds in mechanistic studies (see below). However, potency may not be the only nor the most important element for activation, with the degree of rate enhancement induced by the compound also being an important factor. Understanding the target biology and the degree of activation required to observe the desired phenotypic effect will be important when choosing which compounds to progress. For example, it may be that during any subsequent chemical optimization, increased potency, achieved from improving molecular recognition elements, may be easier to achieve than being able to further improve the rate enhancement of the chemical. Thus, it may be prudent to progress compounds with a high degree of activation even if they are not the most potent hits (see Figure 7.3).

7.4 Addressing the Challenges of Activator Discovery

There are several approaches that can be valuable in overcoming the challenges with activator drug discovery. Firstly, understanding the behavior of the assay system under conditions of activation is important to address early. Mimicking the effects of activation to assess the ability of the assay to perform under the conditions of increased reaction rate is useful. This may require characterizing the behavior of the assay with increased enzyme concentration, increased substrate concentration, or under other conditions that may deliver rate enhancement (temperature, pH). This ensures that the system will be capable of identifying hits that show the expected level of activation.

As activators may reduce the time period during which initial rate conditions occur, kinetic assays are preferred. This may be precluded during primary screening and so understanding the behavior of active compounds may require additional, detailed follow-up in lower throughput experiments before further progression. Balanced assay conditions should be utilized for hit finding. As for inhibitors, this condition, where substrates are used at concentrations close to their K_m values, should provide the best chance for identifying activators with different values of α.

It is beneficial that attempts are made to identify tool compounds that might be useful in establishing initial screening parameters and that can be used as positive controls. Rapid, small screens using a diverse range of compounds supplemented with compounds that may have been identified that have a higher probability of success, for example, those identified via virtual screening can be valuable. Replicates should potentially be considered to reduce sampling error and diminish or allow the detection of human and technical errors.

Full matrix experiments, where substrate concentration and activator concentration are varied in the same experiment should be used to obtain full concentration-response curves, which may be analyzed appropriately. Area under the curve (AUC) plots, see Figure 7.6, should be used to visually exemplify the behavior of compounds having different values of α and β, to allow focus on compounds with the desired profile of α and β values.

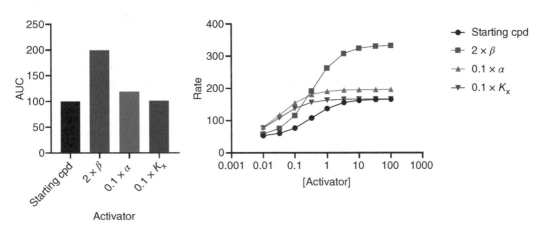

Figure 7.6 Use of area under the curve (AUC) to assess activator effectiveness from concentration-response analyses.
The bar chart (left) shows the area under the curve for four different concentration responses representing activator compounds with different parameter values. The graph on the right shows the activator concentration responses with parameter values described in the legend.

References

1 Gohara DW, Di Cera E. Molecular mechanisms of enzyme activation by monovalent cations. *Journal of Biological Chemistry.* 2016;291(40):20840–8.

2 Baici A. *Kinetics of Enzyme-Modifier Interactions.* Vienna: Springer; 2015.

3 Di Cera E. A structural perspective on enzymes activated by monovalent cations. *Journal of Biological Chemistry.* 2006;281(3):1305–8.

4 Page MJ, Cera ED. Role of Na^+ and K^+ in enzyme function. *Physiological Reviews.* 2006;86(4):1049–92.

5 Toraya T, Sugimoto Y, Tamao Y, Shimizu S, Fukui S. Propanediol dehydratase system. Role of monovalent cations in binding of vitamin B 12 coenzyme or its analogs to apoenzyme. *Biochemistry.* 1971;10(18):3475–84.

6 Shibata N, Masuda J, Tobimatsu T, Toraya T, Suto K, Morimoto Y, et al. A new mode of B12 binding and the direct participation of a potassium ion in enzyme catalysis: X-ray structure of diol dehydratase. *Structure.* 1999;7 8:997–1008.

7 Liao D-I, Dotson G, Turner I, Reiss L, Emptage M. Crystal structure of substrate free form of glycerol dehydratase. *Journal of Inorganic Biochemistry.* 2010;9384–91.

8 Oria-Hernández J, Cabrera N, Pérez-Montfort R, Ramírez-Silva L. Pyruvate kinase revisited: the activating effect of K^+. *Journal of Biological Chemistry.* 2005;280(45):37924–9.

9 Laughlin LT, Reed GH. The monovalent cation requirement of rabbit muscle pyruvate kinase is eliminated by substitution of lysine for glutamate 117. *Archives of Biochemistry and Biophysics.* 1997;348(2):262–7.

10 Larsen TM, Benning MM, Rayment I, Reed GH. Structure of the bis(Mg^{2+})-ATP-oxalate complex of the rabbit muscle pyruvate kinase at 2.1 A resolution: ATP binding over a barrel. *Biochemistry.* 1998;37(18):6247–55.

11 Larsen TM, Benning MM, Wesenberg GE, Rayment I, Reed GH. Ligand-induced domain movement in pyruvate kinase: structure of the enzyme from rabbit muscle with Mg^{2+}, K^+, and L-phospholactate at 2.7 A resolution. *Archives of Biochemistry and Biophysics.* 1997;345(2):199–206.

12 Davis EA, Cuesta-Muñoz A, Raoul M, Buettger C, Sweet I, Moates M, et al. Mutants of glucokinase cause hypoglycaemia- and hyperglycaemia syndromes and their analysis illuminates fundamental quantitative concepts of glucose homeostasis. *Diabetologia.* 1999;42(10):1175–86.

13 Christesen HB, Jacobsen BB, Odili S, Buettger C, Cuesta-Munoz A, Hansen T, et al. The second activating glucokinase mutation (A456V): implications for glucose homeostasis and diabetes therapy. *Diabetes.* 2002;51(4):1240–6.

14 Bonadonna RC, Heise T, Arbet-Engels C, Kapitza C, Avogaro A, Grimsby J, et al. Piragliatin (RO4389620), a novel glucokinase activator, lowers plasma glucose both in the postabsorptive state and after a glucose challenge in patients with type 2 diabetes mellitus: a mechanistic study. *The Journal of Clinical Endocrinology and Metabolism.* 2010;95(11):5028–36.

15 Efanov AM, Barrett DG, Brenner MB, Briggs SL, Delaunois A, Durbin JD, et al. A novel glucokinase activator modulates pancreatic islet and hepatocyte function. *Endocrinology.* 2005;146(9):3696–701.

16 Fyfe MCT, White JR, Taylor A, Chatfield R, Wargent E, Printz RL, et al. Glucokinase activator PSN-GK1 displays enhanced antihyperglycaemic and insulinotropic actions. *Diabetologia.* 2007;50(6):1277.

17 Coope GJ, Atkinson AM, Allott C, McKerrecher D, Johnstone C, Pike KG, et al. Predictive blood glucose lowering efficacy by Glucokinase activators in high fat fed female Zucker rats. *British Journal of Pharmacology.* 2006;149(3):328–35.

18 Grimsby J, Sarabu R, Corbett WL, Haynes NE, Bizzarro FT, Coffey JW, et al. Allosteric activators of glucokinase: potential role in diabetes therapy. *Science.* 2003;301(5631):370–3.

19 Kamata K, Mitsuya M, Nishimura T, Eiki J, Nagata Y. Structural basis for allosteric regulation of the monomeric allosteric enzyme human glucokinase. *Structure.* 2004;12(3):429–38.

20 Zorn JA, Wells JA. Turning enzymes ON with small molecules. *Nature Chemical Biology.* 2010;6(3):179–88.

21 AEvarsson A, Chuang JL, Wynn RM, Turley S, Chuang DT, Hol WG. Crystal structure of human branched-chain alpha-ketoacid dehydrogenase and the molecular basis of multienzyme complex deficiency in maple syrup urine disease. *Structure.* 2000;8(3):277–91.

22 Shimomura Y, Kuntz MJ, Suzuki M, Ozawa T, Harris RA. Monovalent cations and inorganic phosphate alter branched-chain alpha-ketoacid dehydrogenase-kinase activity and inhibitor sensitivity. *Archives of Biochemistry and Biophysics.* 1988;266(1):210–8.

23 Changeux JP. Allostery and the Monod–Wyman–Changeux model after 50 years. *Annual Review of Biophysics.* 2012;41:103–33.

24 Cool B, Zinker B, Chiou W, Kifle L, Cao N, Perham M, et al. Identification and characterization of a small molecule AMPK activator that treats key components of type 2 diabetes and the metabolic syndrome. *Cell Metabolism.* 2006;3(6):403–16.

25 Steinberg GR, Carling D. AMP-activated protein kinase: the current landscape for drug development. *Nature Reviews Drug Discovery.* 2019;18(7):527–51.

26 Anderson SN, Cool BL, Kifle L, Chiou W, Egan DA, Barrett LW, et al. Microarrayed compound screening (microARCS) to identify activators and inhibitors of AMP-activated protein kinase. *Journal of Biomolecular Screening.* 2004;9(2):112–21.

27 Scott JW, van Denderen BJ, Jorgensen SB, Honeyman JE, Steinberg GR, Oakhill JS, et al. Thienopyridone drugs are selective activators of AMP-activated protein kinase beta1-containing complexes. *Chemistry and Biology.* 2008;15(11):1220–30.

28 Turberville A, Semple H, Davies G, Ivanov D, Holdgate GA. A perspective on the discovery of enzyme activators. *SLAS Discovery.* 2022.

29 Kok BP, Ghimire S, Kim W, Chatterjee S, Johns T, Kitamura S, et al. Discovery of small-molecule enzyme activators by activity-based protein profiling. *Nature Chemical Biology.* 2020;16(9):997–1005.

30 Cabrol C, Huzarska MA, Dinolfo C, Rodriguez MC, Reinstatler L, Ni J, et al. Small-molecule activators of insulin-degrading enzyme discovered through high-throughput compound screening. *PLoS One.* 2009;4(4):e5274.

31 Zhang JH, Chung TD, Oldenburg KR. A simple statistical parameter for use in evaluation and validation of high throughput screening assays. *Journal of Biomolecular Screening.* 1999;4(2):67–73.

8

Mechanism of Action

8.1 Introduction

To understand the molecular basis of the action of compounds affecting enzyme activity, it is important to examine the equilibria between enzyme, substrate, and compound that can occur in solution. The usual assumptions about steady-state and the concentrations of compound and substrate relative to enzyme, discussed in Chapter 6 for inhibitors and Chapter 7 for activators, are made for classical inhibitors and non-essential activators. It should be noted that several different physical mechanisms may give rise to a single kinetic mechanism, and so further information may be required to understand the physical details of the interaction producing a specific kinetic mechanism. Details for characterizing both inhibition and activation mechanisms will be given below.

8.2 Mechanisms of Inhibition

Reaction scheme 8.1 shows a general scheme for the potential interactions between an enzyme, E, a single substrate, S, and an inhibitor, I.

In the simplified scheme above, which does not show enzyme-bound intermediates or products, K_m is the Michaelis-Menten constant for the reaction in the absence of I, K_i is the apparent dissociation constant for the EI complex (and reflects potency when $[S] \ll K_m$), αK_i represents the apparent dissociation constant for I from the ES complex (and reflects potency when $[S] \gg K_m$) and k_{cat} is the rate constant for product formation. The term α represents the effect of inhibitor on the affinity of the enzyme for substrate. The term β reflects the modification of the rate of product

Laboratory Guide to Enzymology, First Edition. Geoffrey A. Holdgate, Antonia Turberville, and Alice Lanne.
© 2024 John Wiley & Sons, Inc. Published 2024 by John Wiley & Sons, Inc.

$$E \; + \; S \; \underset{}{\overset{K_m}{\rightleftharpoons}} \; ES \; \xrightarrow{k_{cat}} \; E + P$$

$$+ \qquad\qquad +$$

$$I \qquad\qquad I$$

$$\Big\updownarrow K_i \qquad\qquad \Big\updownarrow \alpha K_i$$

$$EI \; + \; S \; \underset{}{\overset{\alpha K_m}{\rightleftharpoons}} \; EIS \; \xrightarrow{\beta k_{cat}} \; EI + P$$

Reaction scheme 8.1 The equilibria for enzyme turnover in the presence and absence of inhibitor.
This is a general scheme representing partial mixed noncompetitive inhibition from which the schemes for full inhibition following pure noncompetitive, competitive, and uncompetitive inhibition can be derived.

formation caused by the inhibitor. An inhibitor that completely blocks activity will have $\beta = 0$, and a partial inhibitor will have a value of β between 0 and 1.

Most enzyme inhibitors act through one of four different kinetic mechanisms: mixed noncompetitive, pure noncompetitive, competitive, and uncompetitive inhibition. When $\beta = 0$, an association of enzyme and inhibitor completely prevents either substrate binding or catalysis. The different mechanisms are differentiated by the relative affinities of I for free enzyme, E, and the enzyme-substrate complex, ES. Thus, the calculation of the values of K_i and αK_i establishes the mechanism of inhibition (Table 8.1).

8.2.1 Competitive Inhibition

Competitive inhibition refers to the case where the inhibitor potency falls as the concentration of substrate is raised. For example, the binding of the two ligands may be mutually exclusive. This type of inhibition has values of $\alpha = \infty$, so that $\alpha K_i = \infty$ (i.e. binding to ES does not occur) and $\beta = 0$ (complete inhibition is observed, the cases where $\beta \neq 0$ are termed partial inhibition and will be discussed later). In competitive inhibition, the two ligands compete for the same enzyme form, and cannot bind simultaneously (or if I does bind to ES, it does so without exerting any effect on catalytic rate).

Most often, competitive inhibitors function by binding to the active site, thereby directly competing with the substrate for a common site on the enzyme. However, competitive inhibitors are not constrained to bind at the same site as the substrate. It is possible for the inhibitor to bind at a distinct site, leading to a conformational change in the enzyme which modifies the shape of the substrate binding site, preventing substrate binding. These two types of competitive inhibition are shown in Figures 8.1 and 8.2.

Table 8.1 The values of K_i and α for the different mechanisms of inhibition.

Mechanism	I binds to	K_i	α
Mixed noncompetitive	E, ES	Not $= \alpha K_i$	Not $= 1$
Pure noncompetitive	E, ES	$= \alpha K_i$	$= 1$
Competitive	E only	Finite	Infinitely high
Uncompetitive	ES only	Infinitely high	Finite

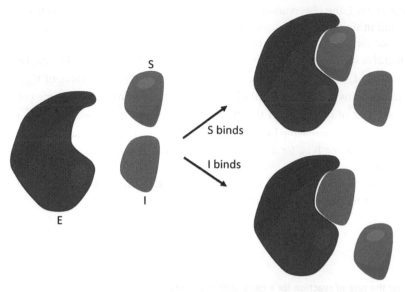

Figure 8.1 Competitive inhibition caused by inhibitor and substrate binding to the same site.
Inhibitor (I) and substrate (S) are mutually exclusive, and so compete for the binding site.

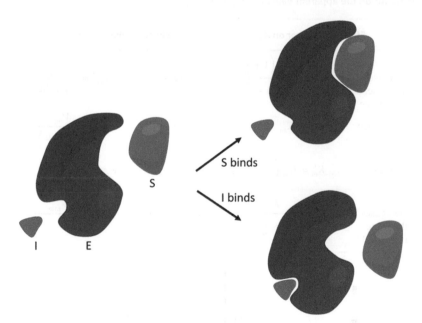

Figure 8.2 Competitive inhibition arising from a conformational change in the binding site of substrate preventing binding of inhibitor and vice versa. Binding of the two ligands is mutually exclusive.
The binding of inhibitor (I) and substrate (S) is mutually exclusive, but the binding sites for each are distinct from each other.

As shown in Figures 8.1 and 8.2, the observation of competitive inhibition does not provide evidence that the substrate and inhibitor use the same site. The best that can be said is that the two ligands compete for the same form of the enzyme, the free enzyme.

The presence of a competitive inhibitor has the effect of apparently raising the K_m of the enzyme for the substrate, by the factor $\left(1 + \frac{[I]}{K_i}\right)$. Competitive inhibition does not affect the value of V_{max}, since infinitely high concentrations of substrate will displace the inhibitor from the enzyme (see Table 8.2). Conversely, the effect of substrate on the observed K_i' is given by $K_i' = K_i \left(1 + \frac{[S]}{K_m}\right)$, hence the value of K_i' increases with increasing [S]. It is important to note that increasing [S] does not always significantly increase K_i' or IC_{50} for a competitive inhibitor. The increase in IC_{50} only becomes significant when $[S] > K_m$.

The full rate equation for competitive inhibition is given by Equation 8.1 and data can be visualized using nonlinear and linear regression analysis (Figure 8.3).

$$v = \frac{V_{max}[S]}{K_m\left(1 + \frac{[I]}{K_i}\right) + [S]}$$

Equation 8.1 **Expression for the rate of reaction for a competitive inhibitor.**

Table 8.2 **The effects of inhibitor on the apparent values of K_m and V_{max}.**

Type of inhibition	Effect of inhibitor on K_m		Effect of inhibitor on V_{max}	
Mixed noncompetitive	Increase, decrease, or no change	$K_m' = K_m \dfrac{\left(1 + \frac{[I]}{K_i}\right)}{\left(1 + \frac{[I]}{\alpha K_i}\right)}$	Decrease	$V_{max}' = \dfrac{V_{max}}{\left(1 + \frac{[I]}{\alpha K_i}\right)}$
Pure noncompetitive	No change	$K_m' = K_m$	Decrease	$V_{max}' = \dfrac{V_{max}}{\left(1 + \frac{[I]}{K_i}\right)}$
Competitive	Increase	$K_m' = K_m\left(1 + \frac{[I]}{K_i}\right)$	No change	$V_{max}' = V_{max}$
Uncompetitive	Decreases by the same factor as V_{max}	$K_m' = \dfrac{K_m}{\left(1 + \frac{[I]}{\alpha K_i}\right)}$	Decreases by the same factor as K_m	$V_{max}' = \dfrac{V_{max}}{\left(1 + \frac{[I]}{\alpha K_i}\right)}$

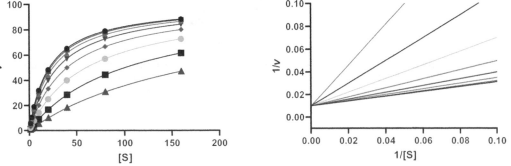

Figure 8.3 **Michaelis–Menten (left) and Lineweaver Burk (right) plots for competitive inhibition.** For competitive inhibitors, the intersection point occurs on the $1/v$ axis at a value of $1/V_{max}$.

Interpretation of competitive kinetics is more complicated for enzymes that use more than one substrate. Consider an enzyme that binds substrates in the order A and then B. An inhibitor could displace A and not allow binding of B. Such a compound would follow competitive kinetics against both substrates, despite using a site for only one of them.

As mentioned above, there are a number of physical mechanisms that may give rise to competitive inhibition. For example, competitive inhibition is observed when I and S compete for a single binding site; when binding of I to an allosteric site on the free enzyme prevents S binding at a different site; and when I binds to both free E and ES complex but binding to free E prevents S binding and binding to ES complex has no effect.

8.2.2 Mixed Noncompetitive and Pure Noncompetitive Inhibition

Mixed noncompetitive inhibition (Figure 8.4) refers to the case in which an inhibitor binds to the enzyme both when $[S] < K_m$ (i.e. to free E) and when $[S] > K_m$ (i.e. to ES complex). This type of inhibition is the general case, and all the other types of inhibition are just special cases, with restricted values of α and K_i.

Two types of noncompetitive inhibition are often described:

Mixed noncompetitive
Pure noncompetitive

In mixed noncompetitive inhibition, the value of α can truly be any finite value, as can K_i. In this case, the inhibitor binds with a different affinity to the free E (when the $[S] < K_m$), compared to the ES complex (or when $[S] > K_m$). Pure noncompetitive inhibition, on the other hand, refers to the case where $\alpha = 1$, and $K_i = \alpha K_i$, so that the inhibitor displays equal affinity to both free E (when $[S] < K_m$) and the ES complex (when $[S] > K_m$). Both forms of noncompetitive inhibition therefore

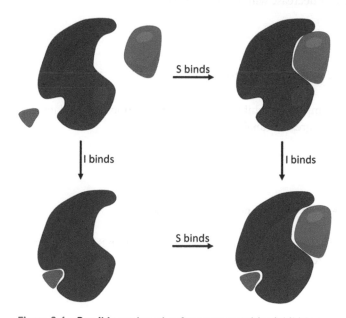

Figure 8.4 Possible explanation for noncompetitive inhibition.
The inhibitor may bind to the free enzyme and also the ES complex. Conversely, the substrate can also bind to both the free enzyme and the EI complex. The relative affinities determine whether inhibition is of the pure form or the mixed form.

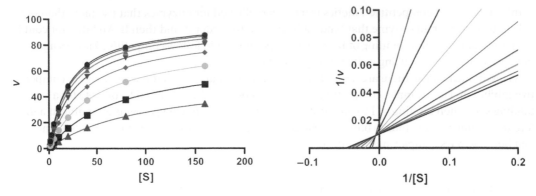

Figure 8.5 **Michaelis-Menten (left) and Lineweaver-Burk (right) plots for mixed noncompetitive inhibition.** For mixed noncompetitive inhibitors, the intersection point of the Lineweaver-Burk plot occurs above the $1/[S]$ axis at a value of $(\alpha - 1)/\alpha V_{max}$.

reflect that inhibitor binding can occur when $[S] < K_m$ or when $[S] > K_m$. The different types of noncompetitive inhibition also have different effects on the measured values of K_m and V_{max}.

For mixed noncompetitive inhibition, the value of K_m increases or decreases according to the factor $\left(\frac{1 + \frac{[I]}{K_i}}{1 + \frac{[I]}{\alpha K_i}} \right)$. The value of V_{max} is decreased by the factor $\left(1 + \frac{[I]}{\alpha K_i} \right)$. As $\alpha = 1$ for pure noncompetitive inhibition, the K_m value for this mechanism of inhibition is unchanged (see Table 8.2), whereas the V_{max} value is decreased by the factor $\left(1 + \frac{[I]}{K_i} \right)$. The effects of increasing substrate concentration on the value of K_i' (Equation 8.2) also depend upon which type of noncompetitive inhibition is followed. As should be clear, for pure noncompetitive inhibitors, the choice of substrate concentration does not affect K_i', since $K_i' = K_i = \alpha K_i (\alpha = 1)$. For mixed noncompetitive inhibition, the value of K_i' can increase or decrease with increasing $[S]$.

$$K_i' = K_i \alpha K_i \frac{([S] + K_m)}{(K_i[S] + \alpha K_i K_m)}$$

Equation 8.2 Expression for K_i' for a mixed noncompetitive inhibitor.

The full rate equation for a pure noncompetitive inhibitor is given by Equation 8.3 and data can be visualized using nonlinear and linear regression analysis (Figure 8.5).

$$v = \frac{V_{max}[S]}{(K_m + [S]) \left(1 + \frac{[I]}{K_i} \right)}$$

Equation 8.3 **Expression for the rate of reaction for a pure noncompetitive inhibitor.**

The full rate equation for a mixed noncompetitive inhibitor is given by Equation 8.4 and data can be visualized using nonlinear and linear regression analysis (Figure 8.6).

$$v = \frac{V_{max}[S]}{K_m \left(1 + \frac{[I]}{K_i} \right) + [S] \left(1 + \frac{[I]}{\alpha K_i} \right)}$$

Equation 8.4 **Expression for the rate of reaction for a mixed noncompetitive inhibitor.**

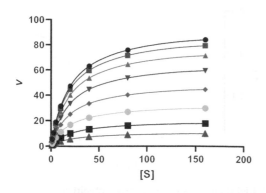
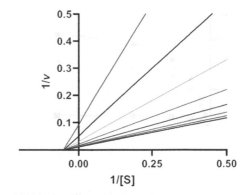

Figure 8.6   Michaelis-Menten (left) and Lineweaver-Burk (right) plots for pure noncompetitive inhibition.
For pure noncompetitive inhibitors, the intersection point occurs on the 1/[S] axis at a value of $-1/K_m$.

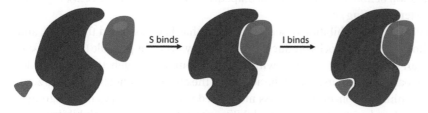

Figure 8.7   Possible explanation for uncompetitive inhibition.
In uncompetitive inhibition, the inhibitor is unable to bind to free enzyme but can bind to the ES complex.

8.2.3   Uncompetitive Inhibition

The binding of uncompetitive inhibitors is favored by increasing [S] relative to K_m (Figure 8.7). Therefore, $\alpha \ll 1$, and $\alpha K_i < K_i$. A true uncompetitive inhibitor has no affinity for free enzyme and hence K_i has a value of infinity, but the inhibitor may bind to ES, and hence the value of αK_i is finite. Often, however, the inhibitor does retain some affinity for free enzyme, but with $K_i \gg \alpha K_i$.

The effect of an uncompetitive inhibitor is to decrease both K_m and V_{max} by the same factor $\left(1 + \frac{[I]}{\alpha K_i}\right)$ (see Table 8.2). The effect of increasing the substrate concentration for uncompetitive inhibitors will lead to more of the enzyme being in a form capable of binding inhibitors, and so the inhibition should be stronger. K_i' decreases with increasing [S], according to Equation 8.5.

$$K_i' = \alpha K_i \left(1 + \frac{K_m}{[S]}\right)$$

Equation 8.5   Expression for the apparent inhibition constant, K_i', for an uncompetitive inhibitor.

The full rate equation for an uncompetitive inhibitor is given by Equation 8.6 and data can be visualized using nonlinear and linear regression analysis (Figure 8.8).

$$v = \frac{V_{max}[S]}{K_m + [S]\left(1 + \frac{[I]}{\alpha K_i}\right)}$$

Equation 8.6   Expression for the rate of reaction for an uncompetitive inhibitor.

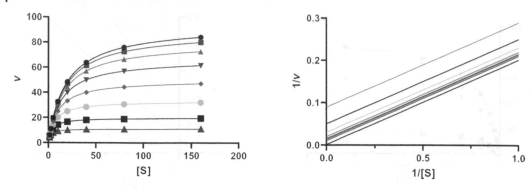

Figure 8.8 **Michaelis-Menten (left) and Lineweaver-Burk (right) plots for uncompetitive inhibition.** For uncompetitive inhibitors, the lines of the Lineweaver Burk plot are parallel, with no intersection.

Note that the observation of noncompetitive or uncompetitive kinetics does not necessarily indicate the use of an allosteric binding site.

For each of the mechanisms of inhibition described above, it has been assumed that $\beta = 0$, and that binding of inhibitor completely blocks product formation. However, it is possible that $0 < \beta < 1$, and that the enzyme can still turn over substrate with inhibitor bound, although at a reduced rate. This is the situation for partial inhibitors, where the activity of the enzyme cannot be driven to zero, even at extremely high inhibitor concentrations. As mentioned above, these types of background rates observed can be confused with compound insolubility, and other causes of uninhibited rates should be explored to determine the true cause of the effect.

8.3 Choosing Between Different Types of Inhibition

Distinguishing between the different mechanisms of inhibition can be achieved by applying nonlinear regression analysis using the equations for the different models (Equations 8.1, 8.3, 8.4, and 8.6), and the techniques described above. This requires data collected from an experiment where both [S] and [I] are varied since it is the effect of the change in [S] on K_i', which determines the mechanism (Table 8.3). Ideally, both [S] and [I] are varied in the same fit of the rate equations to the

Table 8.3 **The effects of changing substrate concentration on the value of K_i' for the various types of inhibition mechanism.**

Type of inhibition	Effect of increasing [S] on K_i'	General formula for K_i'
Mixed noncompetitive	Increases or decreases	$= \dfrac{K_i \alpha K_i([S] + K_m)}{(K_i[S] + \alpha K_i K_m)}$
Pure noncompetitive	None	$K_i = \alpha K_i$
Competitive	Increases	$= K_i \left(1 + \dfrac{[S]}{K_m} \right)$
Uncompetitive	Decreases	$= \alpha K_i \left(1 + \dfrac{K_m}{[S]} \right)$

Knowledge of the mechanism and the substrate concentration then allows the calculation of K_i and/or αK_i from K_i'.

data, in a global or multivariate fit. The results are less informative when separate univariate fits (varying [S] at fixed [I] or varying [I] at fixed [S]) are carried out.

8.4 Interpretation of Mechanism of Inhibition

Caution should always be used when attempting to interpret data collected on the mechanism of inhibition of compounds. For instance, it is often assumed that competitive kinetics show that compounds use a site that overlaps with the substrate site and that noncompetitive or uncompetitive kinetics demonstrate that the compound uses a different site to the substrate. This can, of course, be the case, but there are alternative explanations. Kinetic data provide only indirect information about the identity of the binding site.

To illustrate this point, imagine an assay where compounds block a protein–protein interaction, leading to a decrease in the measured rate. Binding to either partner would lead to the same concentration-response relationship, but no information about to which protein the compound actually binds is gained in these kinetic experiments. Separate binding studies for each partner are required.

Another example that may be encountered is an ordered mechanism for 2-substrate reactions (Reaction scheme 8.2) – many dehydrogenases, reductases and some kinases may follow this mechanism. The determined mechanism of inhibition in this case may be misleading at first sight. In this case, if substrate A binds to the enzyme before substrate B, and inhibitors use the site for A, competitive kinetics with respect to B may be observed. This is because the inhibitor binds before the varied substrate (B in this case). Hence, competitive kinetics with respect to B may suggest that the compound binds at the site for B, when really it binds at the site for A.

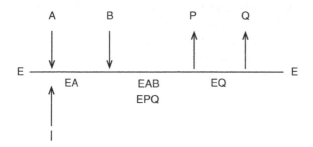

Reaction scheme 8.2 A 2-substrate reaction ordered mechanism.
Reaction scheme, displayed in the Cleland notation, for the ordered bi reaction, where substrates A and B bind in a defined order to the enzyme, E. A binds first then B, with release of products P followed by Q. If an inhibitor binds at the site for A, it may show competitive inhibition with respect to both A and B, even though it does not occupy the site for B.

This illustrates how it is extremely useful to know something about the order of addition of the substrates and order of dissociation of the products for the reaction in the absence of inhibitors. Similarly, it can be useful to know the identity of the rate-limiting step of the enzyme reaction. For example, an inhibitor may use a substrate site after the substrate has bound, been converted to product, and dissociated from the enzyme. This situation will give noncompetitive or uncompetitive kinetics even though the inhibitor binds in the substrate site because the compound binds after the varied substrate (see Section 8.6.3).

A knowledge of the kinetic mechanism in the absence of inhibitors can be of considerable value in recognising these situations and therefore help to avoid misleading interpretations from mechanism of inhibition studies.

Determination of the sequence of substrate addition and identification of the rate-determining step is of considerable value in understanding the molecular mechanism of active compounds. This can increase efficiency and effectiveness in lead optimization. However, this information may not be available. In its absence, it may be useful to characterize inhibition of known analogs of substrates to define a profile, which can be used to understand test compounds. For example, anilinoquinazolines inhibitors of protein kinases exhibit the same profile as ADPNP (ATP – competitive, peptide substrate – noncompetitive) and a different profile to a Tyr → Phe substrate analog (ATP – noncompetitive, peptide substrate – competitive).

Conversely, it is not always possible to predict the kinetic mechanism from the crystal structure of a compound bound to an enzyme, yet it is the kinetics that controls the biological activity of that compound. This illustrates the need to combine high-quality structural information with rigorous kinetic analysis to have any real understanding of the mechanism by which interesting compounds have their effect. It is the combination of multiple techniques used to gather this information that provides a powerful means of understanding and improving chemical leads.

8.5 Effect of Multiple Substrates and Assignment of Mechanism

So far, only single substrate reactions have been considered, but this situation does not reflect many enzyme reactions, which tend not to be uni-reactant. Often, enzyme reactions use two or more substrates and follow ordered sequential mechanisms. Many enzyme inhibitors have been designed as substrate analogs and often display competitive inhibition. However, such compounds do not always display competitive effects. Some reasons for this behavior are given below, and these may lead to misleading conclusions and the values of the dissociation constants for EI and ESI complexes may not be equal to microscopic constants. Accessing the true values is important for reliable comparison of potency and correct assignment of kinetic mechanism.

Consider the simplified system in Reaction scheme 8.3, where inhibitors I and X compete with two substrates A and B.

Reaction scheme 8.3 **The reaction scheme for two inhibitors competing with two substrates.**

The maximum rate in the absence of I and X is given by Equation 8.7.

$$V_{max} = k_{cat}[E]_t$$

Equation 8.7 Expression for the maximum rate of reaction for scheme 8.2 in the absence of inhibitors.

And the rate in the presence of I and X is given by Equation 8.8.

$$v = \frac{V_{max}}{\left(1 + \frac{K_b}{[B]} + \frac{K_a K_b}{[A][B]} + \frac{K_b[X]}{[B]K_x} + \frac{K_a K_b[I]}{[A][B]K_i} + \frac{K_a K_b[I]}{[A]K_i K_{ib}}\right)}$$

Equation 8.8 **Expression for the rate of reaction for scheme 8.2 in the presence of two inhibitors.**

In the absence of both inhibitors, the rate is given by Equation 8.9.

$$v = \frac{V_{max}}{\left(1 + \frac{K_b}{[B]} + \frac{K_a K_b}{[A][B]}\right)}$$

Equation 8.9 **Expression for the rate of reaction for scheme 8.2 in the absence of inhibitors.**

When A is the varied substrate, the observed K_m for A is described by Equation 8.10. When [B] is very low ([B] ≪ K_b) then $K_{ma} = K_a$. At higher concentrations of B, K_{ma} is less than K_a.

$$K_{ma} = \frac{K_a K_b}{K_b + [B]}$$

Equation 8.10 **Expression for K_m when A is the varied substrate.**

At saturating concentrations of A, the value of V_{max} varies according to Equation 8.11. This means that V_{maxa} is equal to V_{max} only when B is saturating ([B] ≫ K_b). At concentrations below saturation, V_{maxa} is less than V_{max}.

$$V_{maxa} = \frac{V_{max}[B]}{(K_b + [B])}$$

Equation 8.11 **Expression for V_{max} when A is the varied substrate.**

When B is the varied substrate, the observed K_m value for B is described by Equation 8.12. When A is saturating (≫ K_a), $K_{mb} = K_b$. At concentrations below saturation, K_{mb} is greater than K_b. When B is saturating, the value of V_{max} is equal to V_{maxb} and so the apparent V_{maxb} is always equal to V_{max}.

$$K_{mb} = \frac{K_b(K_a + [A])}{[A]}$$

Equation 8.12 **Expression for K_m when B is the varied substrate.**

In the presence of inhibitor I only, the observed rate is described by Equation 8.13.

$$v = \frac{V_{max}}{\left(1 + \frac{K_b}{[B]} + \frac{K_a K_b}{[A][B]} + \frac{K_a K_b [I]}{[A][B] K_i} + \frac{K_a K_b [I]}{[A] K_i K_{ib}}\right)}$$

Equation 8.13   Expression for the rate of reaction for scheme 8.2, in the presence of inhibitor I only.

When the concentration of A is varied, the apparent K_m for A is described by Equation 8.14, and the apparent $V_{maxa}' = V_{maxa}$. This demonstrates that inhibition is competitive with respect to A, since K_m is increased, and V_{max} is unchanged.

$$K_{ma}' = K_{ma}\left(1 + \frac{[I]}{K_I}\right)$$

Equation 8.14   Expression for the apparent K_m when A is the varied substrate.

The value of K_I is given by Equation 8.15. This demonstrates that the value of K_I is only equal to the microscopic dissociation constant, K_i, when the concentration of B is much lower than K_{ib}.

$$K_I = \frac{K_i K_{ib}}{(K_{ib} + [B])}$$

Equation 8.15   Expression for K_I when only I is present.

When the concentration of B is varied, the observed K_m value for B is described by Equation 8.16 and the apparent V_{max} is given by Equation 8.17.

$$K_{mb}' = \frac{K_{mb}\left(1 + \frac{[I]}{K_I}\right)}{\left(1 + \frac{[I]}{K_{II}}\right)}$$

Equation 8.16   Expression for the apparent K_m when B is the varied substrate.

$$V_{maxb}' = \frac{V_{maxb}}{\left(1 + \frac{[I]}{K_{II}}\right)}$$

Equation 8.17   Expression for the apparent V_{max} when B is the varied substrate.

Here, the general case for inhibition is mixed noncompetitive, since K_m and V_{max}, are changed to different extents. However, the observed mechanism of inhibition will depend upon the values of K_I and K_{II}, as described in Equation 8.18.

$$K_I = \frac{K_i(K_a + [A])}{K_a}$$

and

$$K_{II} = \frac{[A] K_i K_{ib}}{K_a K_b}$$

Equation 8.18   Expression for the macroscopic inhibition constants K_I and K_{II}.

Hence, the observed mechanism of inhibition changes as shown in Equation 8.19. When A is saturating, there is no inhibition.

The inhibition follows a mixed noncompetitive mechanism, when

$$K_{ib} \neq \frac{(K_a + [A])K_b}{[A]}$$

Pure noncompetitive kinetics are observed when

$$K_{ib} = \frac{(K_a + [A])K_b}{[A]}$$

An uncompetitive mechanism is followed when

$$K_{ib} \ll \frac{(K_a + [A])K_b}{[A]}$$

And competitive kinetics demonstrated when

$$K_{ib} \gg \frac{(K_a + [A])K_b}{[A]}$$

Equation 8.19 **Expressions demonstrating the different mechanisms of inhibition followed depending upon the magnitude of K_{ib}, when B is the varied substrate.**

In the presence of inhibitor X only, the rate equation is shown in Equation 8.20.

$$v = \frac{V_{max}}{\left(1 + \frac{K_b}{[B]} + \frac{K_a K_b}{[A][B]} + \frac{K_b[X]}{[B]K_x}\right)}$$

Equation 8.20 **Expression for the rate of reaction for scheme 8.2, in the presence of inhibitor X only.**

When the concentration of A is varied, the apparent value of K_m is shown by Equation 8.21 and the observed value of V_{max} is shown by Equation 8.22.

$$K_{ma}' = \frac{K_{ma}}{\left(1 + \frac{[X]}{K_{XX}}\right)}$$

Equation 8.21 **Expression for apparent K_m when A is the varied substrate.**

$$V_{maxa}' = \frac{V_{maxa}}{\left(1 + \frac{[X]}{K_{XX}}\right)}$$

Equation 8.22 **Expression for the apparent V_{max} when A is the varied substrate.**

As both K_m and V_{max} are decreased by the same value, the inhibition is uncompetitive, where K_{XX} is described by Equation 8.23.

$$K_{XX} = K_x \left(1 + \frac{[B]}{K_b}\right)$$

Equation 8.23 **Expression for the macroscopic inhibition constant K_{XX}.**

Again, the value of K_{XX} is only equal to the microscopic value, K_x, when the concentration of B is very low ($[B] \ll K_b$). If this is not the case, then K_{XX} is greater than K_x and the degree of inhibition decreases as $[B]$ increases. As with the case above, the values of the apparent and true dissociation constants can be very different. In this case, however, the uncompetitive mechanism is expected as X can only bind after substrate A has been associated with the enzyme.

When the concentration of B is varied, the apparent value for K_m is shown by Equation 8.24 and the observed value of V_{max} is described by Equation 8.25.

$$K_{mb}' = K_{mb}\left(1 + \frac{[X]}{K_X}\right)$$

Equation 8.24 Expression for the apparent K_m when B is the varied substrate.

$$V_{maxb}' = V_{max}$$

Equation 8.25 Expression for the apparent V_{max} when B is the varied substrate.

This means that inhibition is competitive as only the value of K_m is changed. The value of K_X does not need to be the same as the true microscopic dissociation constant, K_x, as can be seen from the relationship in Equation 8.26.

$$K_X = K_x\left(1 + \frac{K_a}{[A]}\right)$$

Equation 8.26 Expression for apparent inhibition constant, K_X, when the concentration of B is varied.

It can be seen that K_X only equals K_x when the concentration of A is saturating ($[A] \gg K_a$). When A is not saturating, then K_X is greater than K_x and so increasing the concentration of A increases the degree of inhibition. Competitive inhibition is expected from the inspection of Reaction scheme 8.3, as X is an analog of B. However, the value of the observed dissociation constant may be misleading.

This system demonstrates that the apparent kinetic parameters may vary in magnitude according to the conditions, with the true equilibrium constants for individual steps often being very different from the observed values. The variability of these parameters means that the observed kinetic mechanism may be different from the physical processes of binding occurring on the enzyme. This is clear from the fact that the inhibition of compound I, which physically competes with substrate A, can follow any kinetic mechanism when the substrate B is varied.

8.6 Binding Site may not Equal Mechanism

There are examples where the occupation of a particular binding site may not lead to the expected mechanism of inhibition. This highlights the fact that both structural information, which is required for molecular design, and kinetic data, which is required to measure and optimize biological activity should be used together to understand the inhibitor binding process and its consequences. It can be risky to try to predict kinetics from structure or vice versa. It is important

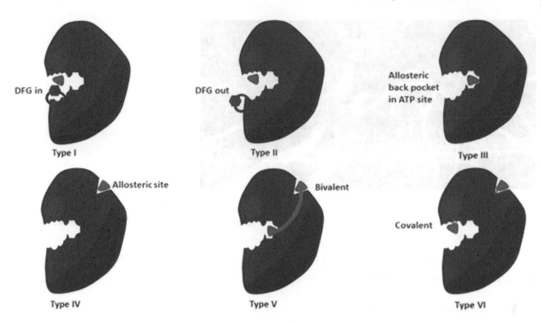

Figure 8.9   Classification of kinase inhibitors depending on the nature of the binding site.
Kinase inhibitors may bind in various binding sites on the kinase protein and have been classified into different types depending upon the binding site and the nature of the interaction and conformational effects on the protein as well as the type of compound. Type I compounds bind to the active conformation of the protein, with the DFG motif on the activation loop occupying the ATP pocket. Type II compounds bind to an inactive conformation of the kinase where the DFG does not occupy the ATP pocket. Type III compounds bind to an allosteric pocket at the back of the ATP site. Type IV compounds bind at a separate allosteric site. Type V compounds interact with both the ATP site and an allosteric site. Type VI compounds form covalent bonds to either the ATP site or an allosteric pocket.

to remember that competitive inhibition may occur at sites remote from the substrate binding site and that conversely, non- or uncompetitive inhibition does not require an allosteric site.

8.6.1   Substrate Competitive Inhibitors at Allosteric Sites

Kinase inhibitors are often classified into different types depending on the nature of the binding site, as shown in Figure 8.9. There are numerous examples of allosteric ATP competitive kinase inhibitors, including inhibitors of p38α [1], FAK [2] and IkB [3]. These compounds do not bind in the ATP site, nor interact with the hinge, but are competitive with ATP and ATP-site probes.

8.6.2   Noncompetitive Binding giving Competitive Inhibition

Figure 8.10 shows that triclosan binds adjacent to the nicotinamide ring of the nucleotide cofactor in the active site of enoyl-acyl carrier protein reductase (EACPR). The phenol ring of triclosan forms a face-to-face interaction with the nicotinamide ring, allowing extensive stacking interactions. Additional van der Waals contacts are made by both rings of triclosan with residues lining the active site and the substrate-binding pocket of the enzyme and with the nucleotide cofactor. Triclosan occupies the site for enoyl ACP, not nicotinamide adenine dinucleotide (NADH) but demonstrates competitive kinetics with respect to NADH. The compound binds only after

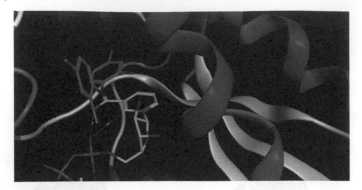

Figure 8.10   The structure of enoyl-acyl carrier protein reductase with triclosan bound.
Triclosan (pink) binds to the active site of EACPR adjacent to the nicotinamide ring of the nucleotide cofactor (green).

NAD^+ and interacts with the bound NAD^+. Reaction scheme 8.4 indicates why the kinetics with respect to NADH are competitive.

$$\pm\,NADH \quad E.NADH \longrightarrow \longrightarrow E + P$$
$$E$$
$$\pm\,NAD^+ \quad E.NAD^+ \rightleftharpoons E.NAD^+.I$$
$$\pm\,I$$

Reaction scheme 8.4   Binding of triclosan to EACPR.
Triclosan binds most tightly to an NAD^+-bound form of EACPR. NAD^+ is a product of the reaction and triclosan binds in an uncompetitive manner when NAD^+ is varied. When NADH is varied, triclosan shows competitive kinetics, even though it does not bind in the NADH binding site but occupies the site for enoyl ACP. This is due to NAD^+ effectively competing with NADH, and NAD^+ being required for the high affinity triclosan complex.

8.6.3   Competitive Binding giving Uncompetitive Inhibition

The structure in Figure 8.11 shows Inosine 5′ monophosphate dehydrogenase (IMPDH) in complex with mycophenolic acid (MPA) and an inosine 5′ monophosphate (IMP) reaction intermediate

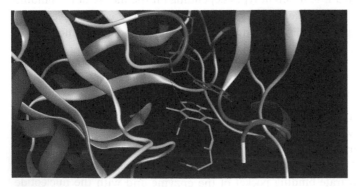

Figure 8.11   The structure of inosine 5′ monophosphate dehydrogenase in complex with mycophenolic acid and inosine 5′ monophosphate.
IMPDH is shown here in a complex with MPA (green) and an IMP reaction intermediate (pink) that is generated during substrate turnover.

that is generated during substrate turnover. In this inhibited state, both hydride transfer and NADH release have occurred, but xanthosine monophosphate (XMP) has not been produced. MPA and the IMP thioimidate intermediate are bound at the active site simultaneously. This observation is consistent with the uncompetitive inhibition of IMPDH by MPA. So, although MPA binds in the NAD$^+$ site, it shows uncompetitive kinetics, as prior binding of NAD$^+$ is required to form the thioimidate intermediate complex to which it binds. This is described in Reaction scheme 8.5.

$$E + IMP + NAD^+ \rightleftharpoons E.IMP.NAD^+ \rightleftharpoons \overset{\pm\,NADH}{E\text{-Int.NADH}} \rightleftharpoons \overset{slow}{E\text{-Int}} \longrightarrow E + XMP$$

$$\updownarrow \begin{array}{c} +H_2O \\ \pm I \end{array}$$

$$E\text{-Int.I}$$

Reaction scheme 8.5 Binding of MPA to IMPDH.
MPA binds to an enzyme intermediate involving a covalent bond between carbon 2 of the purine moiety of IMP and Cys 331. MPA shows uncompetitive inhibition with respect to both IMP and NAD$^+$ substrates but occupies the site for NAD$^+$. This occurs since prior binding of both IMP and NAD$^+$ is required in order to form the thioimidate intermediate to which MPA binds. MPA shows competitive kinetics with respect to XMP consistent with binding to the enzyme intermediate.

8.7 Specificity

Consider an *in vivo* system, in which a single substrate, S, is turned over by several enzymes. If a compound, I, is identified that shows competitive kinetics versus the target enzyme when the concentration of S is varied, the most likely mechanism may be because I is a structural analog of S and utilizes the same binding site. Clearly, this may be problematic as I can act as a substrate analog not only for the target enzyme but for other enzymes that turn over S. This may lead to inhibition of other enzymes and can be associated with unwanted side effects. For example, some kinase inhibitors may lack selectivity due to similarities in the kinase active site, and some lipoxygenase inhibitors may also affect cyclooxygenase. However, in some cases, the inhibition of additional enzymes may be beneficial, for example, some competitive OGG1 inhibitors show important off-target effects by directly inhibiting efflux pumps, disturbing mitotic progression [4].

Noncompetitive and uncompetitive inhibitors are often less likely to resemble the structure of the substrate and so have a lower probability for off-target effects by binding at similar substrate binding sites on additional enzymes. However, these compounds may exert their effects by binding to allosteric sites, which may also be present in other enzymes.

8.7.1 Effect of Increasing [S]

As can be seen in Section 8.5, the effect of substrate concentration varies according to the different mechanisms of inhibition. *In vivo*, the inhibition of an enzyme may have a relatively small effect on the concentration of substrate, since alternative pathways may exist that can regulate concentration changes, and which are less affected by the introduction of inhibitor. If an effect is observed, it is most often an increase in substrate concentration, as inhibition of the enzyme leads to lower turnover. Clearly, an increase in substrate concentration influences compounds with the various mechanisms of inhibition in different ways. The potency of a competitive inhibitor will be decreased. Pure noncompetitive inhibitors are not affected. Uncompetitive compounds become more potent, whilst mixed noncompetitive inhibitors will tend towards the behavior of either competitive or uncompetitive. Thus, it is important to consider the desired mechanism of inhibition when inhibition of the target enzyme leads to significant accumulation of substrate following treatment with compound. Of course, these effects need to be considered in the context of the change in substrate concentration and the free concentration of compound in the vicinity of the target enzyme. For example, if the concentration of free competitive inhibitor is very high relative to its K_i, then a relatively small increase in the concentration of competing substrate may be insufficient to effectively displace the inhibitor.

8.7.2 Effect of [I]

As can be seen from Table 8.2, the observed values of K_m and V_{max} are changed in different ways according to the mechanism of inhibition when the concentration of inhibitor is varied. The effect on V_{max} is a measure of the influence on catalysis, whilst the effect on K_m is inversely related to the effect on substrate binding. An increase in K_m represents a decrease in the apparent affinity for substrate.

It is useful to consider the concentration of inhibitor required to produce the desired decrease in rate of reaction. For example, if the normal physiological concentration of substrate is equal to K_m, and a reduction in rate to 10% (that is 90% inhibition) of the normal level is required for a therapeutic effect, and this degree of inhibition leads to a 10-fold accumulation of substrate, then the concentration of substrate becomes, $[S] = 10 \times K_m$. This situation is expected to have different effects on each mechanism of inhibition.

Considering the initial situation where $[S] = K_m$, where the rate, $v = \frac{V_{max}}{2}$ and the desired decrease in rate is 10-fold. This leads to an inhibited rate of $\frac{V_{max}}{20}$. This can be used to model the inhibitor concentration required for achieving this decrease in rate under two different situations – when substrate does not accumulate (Equation 8.27) and when the accumulation reaches 10-fold, as described above (Equation 8.28).

$$\frac{v_0}{v_i} = 10 = 1 + \frac{[I]}{K_i'}$$

Equation 8.27 **Expression for the relationship between K_i', when the substrate does not accumulate.**

$$\frac{v_0}{v_i} = \frac{200}{11} = 1 + \frac{[I]}{K_i'}$$

Equation 8.28 **Expression for the relationship between K_i', when there is a 10-fold accumulation of substrate.**

Equations 8.27 and 8.11 can be used to estimate the concentration of inhibitor required for each mechanism, as shown in Table 8.4.

To compare the effects of different mechanisms, the inhibitor requirements may be compared to those for the case of pure noncompetitive kinetics, as shown in Table 8.5.

This demonstrates that, when substrate does not accumulate, both competitive and uncompetitive compounds have the same effect on rate and both mechanisms are half as effective as a pure noncompetitive inhibitor. The situation changes slightly when substrate accumulates 10-fold, as the efficacy of a pure noncompetitive compound is 1.1-fold better than an uncompetitive inhibitor and 11-fold better than a competitive compound. It is expected that in practice, most cases of *in vivo* inhibition will fall between these two extreme situations. Thus, this demonstrates that a competitive inhibitor does not suffer from large drops in efficacy compared to other mechanisms

Table 8.4 **The concentration of inhibitor required for each mechanism of action.**

Mechanism	Inhibitor concentration required for 90% decrease in rate	
	When substrate does not accumulate	When a 10-fold accumulation of substrate occurs
Mixed noncompetitive	$\dfrac{18K_i\alpha K_i}{(K_i + \alpha K_i)}$	$\dfrac{189K_i\alpha K_i}{(10K_i + \alpha K_i)}$
Pure noncompetitive	$9K_i$	$17.2K_i$
Competitive	$18K_i$	$189K_i$
Uncompetitive	$18\alpha K_i$	$18.9\alpha K_i$

Table 8.5 **Inhibitor concentration, relative to that of a pure noncompetitive inhibitor, required to bring about 90% inhibition, in the presence and absence of substrate accumulation.**

Mechanism	Inhibitor concentration required for 90% decrease in rate	
	When substrate does not accumulate	When substrate accumulates 10-fold
Competitive	$2\times$ pure noncompetitive	$11\times$ pure noncompetitive
Uncompetitive	$2\times$ pure noncompetitive	$1.1\times$ pure noncompetitive

of inhibition. Hence, a competitive inhibitor with a 2- to 5-fold higher potency may be enough to provide comparable efficacy compared to compounds utilizing other mechanisms. Indeed, if the substrate concentration is below K_m, a competitive inhibitor will be more effective than an uncompetitive compound having the same dissociation constant.

8.8 Activation Mechanisms and Comparison with Inhibition

As mentioned in Chapter 7, non-essential activator binding preferentially to either the free enzyme or to the enzyme-substrate complex leads to different mechanisms of activation, determined by the values of α and β (the factor representing the overall rate enhancement of the catalytic step) [5], shown Reaction scheme 8.6. This general modifier scheme [6] describes the cases

Reaction scheme 8.6 **General modifier scheme.**
The general modifier scheme aims to capture the effects of both inhibitors and activators, where X is the modifier (either inhibitor or activator), α is the reciprocal allosteric coupling factor and β is the effect on k_{cat} (where $\beta > 1$ represents activation and $\beta < 1$, represents inhibition.

of non-essential activation as well as linear and hyperbolic inhibition. The associated general modifier equation (Equation 8.29) can be used to calculate α and β values and therefore define mechanism of action [7].

$$v = \frac{k_{cat}\left(1 + \beta\frac{[X]}{\alpha K_x}\right)[E]_t[S]}{K_s\left(1 + \frac{[X]}{K_x}\right) + [S]\left(1 + \frac{[X]}{\alpha K_x}\right)}$$

Equation 8.29 General modifier equation.

The notion of "competitive" activation can be considered analogous to competitive inhibition, where the activator would compete with the substrate for binding to the free enzyme. However, it clearly cannot occur when $\alpha = \infty$, as this would lead to a dead-end EA complex (in Reaction scheme 8.6), which would be catalytically inert.

"Mixed noncompetitive" activation is similar to mixed noncompetitive inhibition, where the binding of activator may occur with varying affinity before or after substrate has bound. Mixed

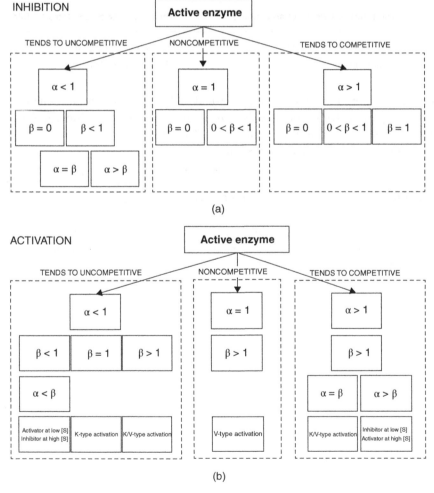

Figure 8.12 Characteristics of enzyme inhibition and activation.

activation where the value of α is very large will tend to disfavor the binding of substrate, and so to demonstrate significant activation the value of β would need to also be large. Compounds conforming to a "mixed" mechanism may achieve activation by various combinations of α and β. For example, compounds may increase the binding affinity of the substrate ($\alpha < 1$, tending towards uncompetitive behavior) or may even decrease the substrate binding affinity ($\alpha > 1$), but will always enhance the rate of the chemical step ($\beta > 1$), see Figure 8.12. In the case of $\alpha < 1$, the substrate binds more tightly to the activator-bound form of the enzyme, leading to higher levels of activation being observed at low substrate concentrations [8–10]. An example is aspartate transcarbamylase (ATCase) which has endogenous positive and negative modulators. Positive modulators bind to the active state, which has higher affinity for its substrate, and small molecules can be designed to mimic these modulating ligands [8].

"Uncompetitive" activation arises when the compound binds only after the substrate has already bound ($0 < \alpha < 1$). This activator mechanism leads to a decrease in apparent K_m and increase in k_{cat} ($\beta > 1$). Activators of this type are exemplified for glucokinase [11, 12]. Examples of compounds that mimic the behavior of a protein binding partner, which subsequently leads to activation of the enzyme, have also been demonstrated for PDK1 [13, 14].

"Pure noncompetitive" activation occurs when the activator can bind, before or after the substrate has bound, with equal affinity to the free enzyme and ES forms. Compounds that bind in this way, where $\alpha = 1$, do not change the apparent K_m but will increase the maximal rate ($\beta > 1$).

Inspection of the general modifier scheme (Reaction scheme 8.6) shows that there are combinations of α and β that can lead to either activation or inhibition depending upon the concentrations of modifier molecule and substrate. This type of behavior may be less common as these mechanisms depend upon the individual values of α and β, but also upon the relative values of α and β. Where $1 < \alpha < \infty$, $1 < \beta < \infty$ and $\alpha > \beta$, then the compound is inhibitory at low substrate concentrations but shows activation at high substrate concentrations. The converse occurs where $0 < \alpha < 1$, $0 < \beta < 1$, and $\alpha < \beta$ [15, 16]. The different mechanisms of inhibition and activation determined by their α and β values and described above are summarised in Figure 8.12.

References

1 Pargellis C, Tong L, Churchill L, Cirillo PF, Gilmore T, Graham AG, et al. Inhibition of p38 MAP kinase by utilizing a novel allosteric binding site. *Nature Structural Biology*. 2002;9(4):268–72.

2 Iwatani M, Iwata H, Okabe A, Skene RJ, Tomita N, Hayashi Y, et al. Discovery and characterization of novel allosteric FAK inhibitors. *European Journal of Medicinal Chemistry*. 2013;61:49–60.

3 Burke JR, Pattoli MA, Gregor KR, Brassil PJ, MacMaster JF, McIntyre KW, et al. BMS-345541 is a highly selective inhibitor of IκB kinase that binds at an allosteric site of the enzyme and blocks NF-κB-dependent transcription in mice. *Journal of Biological Chemistry*. 2003;278(3):1450–6.

4 Tanushi X, Pinna G, Vandamme M, Siberchicot C, D'Augustin O, Di Guilmi A-M, et al. OGG1 competitive inhibitors show important off-target effects by directly inhibiting efflux pumps and disturbing mitotic progression. *Frontiers in Cell and Developmental Biology*. 2023;11:72.

5 Baici A. *Kinetics of Enzyme-Modifier Interactions*. Vienna: Springer; 2015.

6 Di Cera E. A structural perspective on enzymes activated by monovalent cations. *Journal of Biological Chemistry*. 2006;281(3):1305–8.

7 Botts J, Morales M. Analytical description of the effects of modifiers and of enzyme multi-valency upon the steady state catalyzed reaction rate. *Transactions of the Faraday Society*. 1953;49:696–707.

8 Bhagavan NV. Medicinal Biochemistry. In: Bhagavan NV, editor. *Medical Biochemistry* (Fourth Edition). San Diego: Academic Press; 2002. p. 114–9.

9 Grover AK. Use of allosteric targets in the discovery of safer drugs. *Medical Principles and Practice*. 2013;22(5):418–26.

10 Changeux JP. Allostery and the Monod-Wyman-Changeux model after 50 years. *Annual Reviews of Biophysics*. 2012;41:103–33.

11 Grimsby J, Sarabu R, Corbett WL, Haynes NE, Bizzarro FT, Coffey JW, et al. Allosteric activators of glucokinase: potential role in diabetes therapy. *Science*. 2003;301(5631):370–3.

12 Kamata K, Mitsuya M, Nishimura T, Eiki J, Nagata Y. Structural basis for allosteric regulation of the monomeric allosteric enzyme human glucokinase. *Structure*. 2004;12(3):429–38.

13 Engel M, Hindie V, Lopez-Garcia LA, Stroba A, Schaeffer F, Adrian I, et al. Allosteric activation of the protein kinase PDK1 with low molecular weight compounds. *EMBO Journal*. 2006;25(23):5469–80.

14 Balendran A, Biondi RM, Cheung PC, Casamayor A, Deak M, Alessi DR. A 3-phosphoinositide-dependent protein kinase-1 (PDK1) docking site is required for the phosphorylation of protein kinase Czeta (PKCzeta) and PKC-related kinase 2 by PDK1. *Journal of Biological Chemistry*. 2000;275(27):20806–13.

15 Fischer RS, Rubin JL, Gaines CG, Jensen RA. Glyphosate sensitivity of 5-enol-pyruvylshikimate-3-phosphate synthase from *Bacillus subtilis* depends upon state of activation induced by monovalent cations. *Archives of Biochemistry and Biophysics*. 1987;256(1):325–34.

16 Masson P, Froment MT, Gillon E, Nachon F, Lockridge O, Schopfer LM. Kinetic analysis of effector modulation of butyrylcholinesterase-catalysed hydrolysis of acetanilides and homologous esters. *FEBS Journal*. 2008;275(10):2617–31.

9

Data Analysis

9.1 Introduction

Analysis of experimental results is focussed upon the determination of the relationship between the measured values (termed the dependent variable) and a parameter that is knowingly varied (termed the independent variable). There are several methods that may be used to fit models to the generated data, using linear or nonlinear regression analysis. Statistical tests may then be applied to determine the confidence that the model adequately describes the data.

9.2 Statistical Analysis of Enzyme Kinetic Data

The purpose of analysis of enzyme kinetic data is often 2-fold. The first is to identify the correct rate equation to describe the data, and the second is to estimate the values of the parameters contained within that rate equation. The first goal, model discrimination, arises from the fact that several different theoretical models, leading to different rate equations, may be derived, which could be considered to fit the data. The data analysis that is carried out should then help to identify the rate equation that best describes the data collected. Once the best model has been found, it is the estimated values of the parameter values described within that model that are of interest, as well as the error associated with these calculated values.

9.3 Least Squares Fitting

The object of least squares fitting is to determine the best fit of an equation to the data by minimizing the sum of squares of the deviations of the data points from the fitted line (Figure 9.1).

Laboratory Guide to Enzymology, First Edition. Geoffrey A. Holdgate, Antonia Turberville, and Alice Lanne.
© 2024 John Wiley & Sons, Inc. Published 2024 by John Wiley & Sons, Inc.

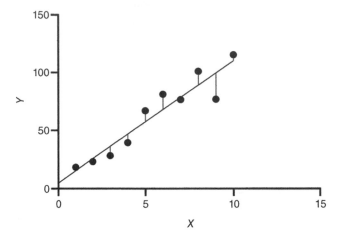

Figure 9.1 Concept for least squares fitting.
Least square fitting of linear plot, red lines show deviations of the data points from the line. The deviations are used to calculate the sums of squares.

Several assumptions are applied when using least squares fitting: the error in the experimental data is described by a normal distribution around the true value; the uncertainty exists only in the dependent variable; the relationship between experimental error and the value of the dependent variable is known, so that the data may be weighted accordingly.

By taking the squares of the residual differences, the fitted line is heavily dependent upon the values of the data points which are the furthest from the best-fit line. If the fitted model is correct, these data points are the least reliable. The issue is that the best-fit line is influenced more by unreliable points than those which are closer to the line so that least squares fitting is described as not being robust, as the presence of outliers in the data are likely to result in a poor fit. Procedures to correct this problem have been proposed [1, 2], describing robust weighting, which applies a weighting function to reduce the influence of outliers.

Many biological phenomena, including many in enzyme kinetics, however, do not follow a linear relationship between the dependent variable and the independent variable. Due to the difficulty of accurately deriving parameter values from curved plots and the lack of computer-based fitting, historically, an algebraic rearrangement of the rate equations was undertaken to obtain a linear plot (e.g. Lineweaver-Burk plots in enzyme inhibition). The relevant parameters were then obtained from the determined values of the gradient and intercept.

Although this approach allows correct estimation of parameter values, providing that correct weighting is used, when the measured value is different from the true value, the error in the weighting factor is amplified accordingly. Thus, it is now preferable to fit directly into the nonlinear relationship, where weighting is generally not required.

9.4 Nonlinear Regression

To avoid issues arising from changing the error structure (which describes how the errors in the data are distributed) during transformation, data for the dependent variable should be analyzed minimizing any manipulation of the data. Thus, mathematical operations, such as calculating

% inhibition or taking ratios or reciprocals, should be avoided. Mathematical operations on the independent variable generally do not cause issues, as this variable is usually relatively free from error, compared to the dependent variable. This lack of manipulation generally means that complex error distributions (errors arise from not knowing the parameter values, which is an unavoidable consequence of having fewer than an infinite number of measurements, and complex error distributions are those having distributions that often are not Gaussian) are not experienced during nonlinear regression analysis. Thus, nonlinear fitting greatly enhances the accuracy of the calculated parameter values and additionally provides for the estimation of standard errors of those parameters, the magnitude of which provides a measure of confidence in the calculated parameter value [1–3].

Nonlinear regression is the method of choice for fitting a model, which describes the observed or dependent variable (y) as a function of the independent variable (x), according to a relationship described by a nonlinear function, $y = f(x)$, to the data.

There are certain assumptions that must be fulfilled before nonlinear regression may be undertaken: the independent variable (x, e.g. often [S], [I], time, etc.) is associated with no error, care should be used, for example, when carrying out dilutions that the actual final concentrations are used for the data analysis, or that time points are accurately recorded.

The variation in the dependent variable (y, e.g. often rate, [P], etc.) follows a normal (Gaussian) distribution. This is usually the case, but other distributions are possible. For unweighted nonlinear regression, the data are assumed to be homoscedastic. This means that the amount of scatter (or lack of precision) is the same for each point on the curve. For weighted nonlinear regression, the relationship of the scatter relative to the y value must be known, so that the correct weighting can then be applied. The observations are independent; a y value at any particular value of x is not dependent on any other measured y value at that value of x.

The data used for nonlinear regression should ideally be the raw data, corrected for any blank rates or background effects. It is usually best not to transform the data in any other way, such as calculating % inhibition or 1/rate. Replicates should be entered as they stand and should not be averaged. Once the data have been collected and the assumptions above are known to be valid, the next step is to fit a model or several models to the data. It is usually best practice to start with the simplest equation, derived from a physical model proposed to describe the variation of y with x. More complex models may be fitted if this simple model is not adequate to describe the data.

To begin the nonlinear regression process, initial estimates for the parameter values need to be provided. Some programs, for some models, may be able to provide these initial estimates automatically, whilst others require that these are input by the user. The quality of the fit is assessed using these initial start points by calculating the residual sum of squares before the variables are changed and the quality of fit is then assessed again (Figure 9.2). The process by which the program adjusts the initial values to improve the fit is called iteration. This process is repeated for several iterations until the best fit is found, which corresponds to the minimum sum of squares. There are several algorithms (e.g. Gauss-Newton algorithm, Gradient descent algorithm, or Levenberg-Marquardt algorithm) that may be applied to ensure that the true minimum has been obtained, rather than just a low value of the sum of squares. Caution must be employed to ensure that these computational methods do not identify a set of parameters that have a lower sum of squares than many other values (described as a local minimum), rather than the true best fit, defined by the lowest possible sum of squares (called the global minimum). Convergence is reached when the sum of squares cannot be further reduced. Various software packages can be employed for nonlinear regression analysis, and these programs allow the application of

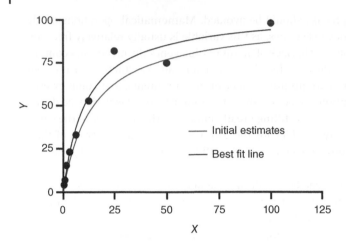

Figure 9.2 Plot showing the difference between the curve generated from initial parameter estimates versus that generated from the best-fit parameter estimates.
When beginning nonlinear regression, initial estimates for the parameter values need to be provided, which can be calculated by the curve fitting program or inputted by the user, and the residual sum of squares is calculated. Iterations are performed to optimize curve fitting by changing the initial estimates for the parameters and gaining the best fit corresponding to a low value of the sum of squares.

weighting, for example, robust weighting to allow for outliers, and provide not only the fitted parameter values and their standard errors but also provide residual plots and the value of the residual sum of squares or a parameter, χ^2 (reduced chi-squared), which is related to the residual sum of squares (SS_{res}) (Equation 9.1).

$$SS_{res} = \frac{\chi^2}{(N - p)}$$

Equation 9.1 Expression for the residual sum of squares.
where χ^2 is the reduced chi-squared value, N is the number of data points and p is the number of parameters.

Many programs also allow the use of more than one independent variable, which is particularly valuable for analyzing enzyme kinetic data, where, for example, inhibitor and substrate concentrations may be varied in a single experiment, generating a single data set.

Key Concept: Steps for Preparing Data for Nonlinear Regression

- Do not transform nonlinear data into linear data and do not smooth the data.
  Note that transforming x-values does not affect the parameter values or their standard errors but that transforming the y-values has the effect of producing different, but equivalent parameter values and standard errors when using linear transformations and different parameter values and standard errors when nonlinear transformations are employed. Nonlinear transformations may be appropriate when the y values at a given x value are not normally distributed.
- It is good practice, to ensure rapid computation, to keep x and y values between 1×10^{-9} and 1×10^9.

- It is useful to include all replicated values in the analysis. There is no need to take averages.
- Caution should be used when considering removing outliers. This can be done where these points are known not to be part of the normal distribution of data.
- Choose the simplest, sensible model that has a physical meaning in relationship to the data as the first model to try.
- Apply the relevant weighting to the data if required.
- Start with initial estimates that you believe may be close to the real values – this will help the model to converge on the global minimum residual sum of squares and not a local minimum.
- If the data do not describe all the parameters in the model, one or more parameters may be constrained to a constant value if these are known accurately.

Once the program has completed its task, by reaching convergence and generating parameter values, it is then up to the user to decide if the results of the nonlinear regression are meaningful. The nonlinear regression program uses a mathematical approach to achieve its goal, it has no understanding of the system under study and cannot tell if the output parameter values make sense. For example, it is not correct to accept values for parameter values that cannot make physical sense, such as negative values for rate constants or, equilibrium constants or concentrations, or parameter values that are beyond physical limits (e.g. rate constants faster than the diffusion-controlled limit).

Sometimes software packages erroneously report incorrect estimates due to finding a local minimum sum of squares, rather than the desired global minimum, as mentioned above. This is often more common for complex models (e.g., where both [S] and [I] are varied). It can be avoided by inspecting the curves generated from the initial estimates, and by testing whether different initial estimates generate the same best fit values. It is, therefore, important that the results of the nonlinear regression analysis should always be investigated carefully.

The fitted curve should be inspected to ensure that it accurately reflects the data – the curve should go through the data points. The value of R^2 (the coefficient of determination) can provide an indication that the curve lies near the data points (Equation 9.2).

$$R^2 = 1 - \frac{SS_{res}}{SS_{tot}}$$

Equation 9.2 Expression for the coefficient of determination.

The best-fit values should make sense and should have small standard errors or narrow 95% confidence intervals, so that the certainty in the best-fit values is high. If the fitted parameter values do not make sense, it may be possible to constrain them to sensible values. If the parameter values are not precise, consider repeating the experiment. The standard errors are calculated from one set of data based on the scatter of the points around the curve. Repetition will demonstrate how the parameter varies from experiment to experiment. This includes more sources of experimental variation than observed in a single experiment.

The residual plot (this is a plot of the difference between the data points and the fitted curve (see below)) should be inspected, to ensure that the residuals are small and randomly distributed. Issues with the fit may be indicated if the residuals are large and systematic. If there are issues with the fitted curve, the parameter values do not make physical sense, or the curve does not adequately describe the data, without producing large, non-random residuals, then consider whether another model is more appropriate.

If poor fits are obtained, there are some key troubleshooting steps that can be followed to improve the fitting process. Firstly, it is simple to try different initial estimates to ensure that the fit has converged on the global minimum. It also may be possible to obtain a similar quality of fit by removing redundant parameters, or by simplifying the model. Ambiguous fits may be avoided by sharing or constraining parameters if the values are known or can be otherwise estimated.

Key Concept: Assessing Output Following Nonlinear Regression

Nonlinear regression analysis is a powerful method for determining parameters from a known model or for determining the best model from several potential options. However, the algorithm just minimizes the distance from the data points and the fitted line. It does not "know" whether the parameter values make physical sense and whether they are reasonable given other observations about the system under study. Hence, it is valuable to interact with the output to ensure that the correct conclusions are drawn.

Where possible, the fitted curve should be observed to ensure that this adequately describes the data. The residuals should be inspected and are expected to be small and randomly distributed. If they do not appear to be, then a runs test may be carried out. The parameter values should be checked – do they make sense – negative values of rate constants or concentrations are not allowed. Parameters with wide confidence intervals should be treated with caution – are there ways of assessing the value more precisely? Check to see if the parameter values correlate. This may suggest that the data are not able to reliably estimate all the parameters in the model. The value of the sum of squares or R squared should be checked and tested for convergence on a local minimum by modifying the initial estimates to see if the same convergence is reached.

Applying this level of rigor in assessing the output from nonlinear regression helps to avoid erroneous values being generated and stored in databases.

9.5 Weighting of Experimental Data

The accuracy of experimental measurements can vary within a data set. If this occurs, then the individual data points should be weighted according to the accuracy in their measurement, to obtain the best fit. The experimental error may have a simple or complex distribution when the residuals between the observed and true values are plotted as a function of the dependent variable. For a simple distribution, the residuals remain fairly constant with increasing magnitude of the independent variable (constant error). However, a complex distribution is observed when the residuals change according to the value of the independent variable (proportional error), see Figure 9.3. For nonlinear regression, weighting does not need to be applied if the experimental error is constant. However, if the error is complex, then the appropriate weighting should be applied during fitting.

To determine the correct weighting, the residuals between the observed and the true value of the dependent variable should be plotted against the true value of the dependent variable. However, most data sets are too small to allow reliable identification of the correct type of weighting that should be used. Estimates have suggested that at least five replicates for each data point are required to obtain a reliable description of the error distribution [4]. In the absence of many replicates, there is no method of weighting that is appropriate for all small data sets.

It has been suggested that the experimental error in enzyme kinetics tends towards being proportional to the measured value [4, 5], although caution should be exercised as low values of rate

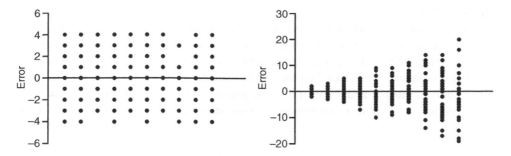

Figure 9.3 Constant error and proportional error.
Constant error (left) shows the error around the data points remains consistent, the error does not change between smaller and larger magnitudes. Proportional error (right) shows the error increasing across the data points, small magnitudes will have small errors and larger magnitudes will have larger errors.

are difficult to measure, especially when the non-enzymatic background is of similar magnitude. This issue of low signal-to-noise leads to a simple error distribution at low values of the dependent variable. As such, a general rule of thumb is that when measurements cover a range of only around 5-fold in signal, then the error may be assumed to be approximately constant, and no weighting should be required during nonlinear regression. When measurements cover a range of 10-fold or more, then the error tends towards being proportional to rate, so that proportional weighting should be used for nonlinear regression.

It is important to remember that evaluating curve fitting involves testing the model and the weighting, so it is vital that the same weighting is used when comparing models fitted to the same data set. The type of weighting applied should also be reported when communicating the results of data analysis procedures. Note that robust weighting, to minimize the effect of outliers, should only be used when the correct model is known, and estimation of the parameter values is the goal. Robust weighting should not be used when comparing models or evaluating a single model.

9.6 Evaluation of Potential Different Models

Once a set of experimental data have been collected, it is often appropriate to decide which mathematical model should be chosen for the analysis of the data. The choice of the correct equation is important as fitting to an inappropriate model usually provides misleading conclusions and incorrect parameter values. Whilst all the available scientific information should be used to aid in the choice of model, there are several qualitative and quantitative methods that can be applied to assess the alternative models.

9.6.1 Distribution of Residuals

Whereas a plot of the residual against the dependent variable illustrates how the data should be weighted, a plot of the residual versus the independent variable provides an indication of the suitability of the model employed in the data analysis. It is expected that an approximately normal distribution of the residuals with respect to the independent variable would be consistent with the differences being produced solely from experimental error [6, 7]. Systematic distribution of residuals often indicates that data are being analyzed by an equation that does not adequately describe the system.

9.6.1.1 The Runs Test

The runs test interrogates the residuals to ascertain whether the fitted curve deviates from the data in a systematic manner, see Figure 9.4. A run is defined as a series of successive points that are all above or below the fitted curve. The runs test compares the count of the actual number of runs with the predicted number of runs, based on the number of data points analyzed. To use the runs test, certain conditions must be applicable: it must be possible to classify each residual value into one of two categories, clearly, this is possible as the residual values are expected to lie above or below the fitted curve. The test will be 2-sided as too few runs suggest there is not enough variation and that the number of runs would occur from a random process. Too many runs occur when the process alternates between categories too frequently to be described by chance. If the curve does not adequately describe the data, then there is expected to be a clustering of points above and below the fit line. The null hypothesis is that the sequence of runs is random. If the runs test reports a low p value (less than the significance level – e.g. <0.05 at the 95% significance level), then the conclusion is that the curve does not describe the data well.

9.6.2 Magnitude of Standard Errors

A mathematical model that does not adequately describe the experimental data produces relatively large standard errors for the fitted parameters. A large standard error may be considered as one that is greater than around 25% of the calculated parameter value [1].

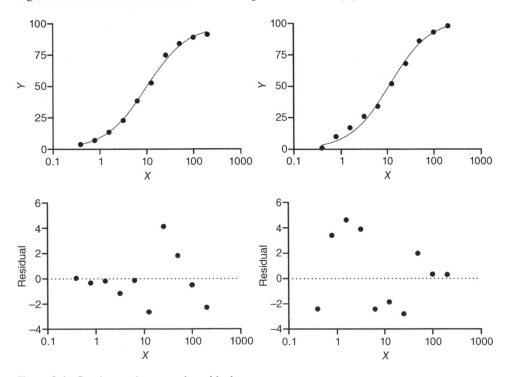

Figure 9.4 Random and systematic residuals.
Random residuals (left) indicate that the data have been fitted to the correct model. Systematic residuals (right) indicate that the data have not been fitted to the correct model. A runs test can be performed to determine whether the residuals are random or systematic, it measures the number of points above and below the fitted line, if this occurs too frequently then the error will be defined as systematic, and the curve fitting model is incorrect.

When the magnitude of an independent variable is changed, the calculated values for the parameters should overlap, within the range of the calculated standard errors. Failure of the parameter values to overlap when the standard errors are taken into account suggests that an incorrect model is used. The probability of the true value falling within a range of ± 1 standard error of the calculated parameter value is around 68%. Considering two standard errors increase the probability to around 95%.

9.6.3 Quantitative Evaluation

The qualitative methods described above (Sections 9.6.1 and 9.6.2) do not provide a numerical measure of the confidence one may have in a particular model. As a result, they do not give a rigorous evaluation of the model. Hence, quantitative methods are required to provide a more conclusive assessment. There are two commonly applied methods for comparing models: the variance ratio or F-test and the Akaike information criterion (AIC) approach based on information theory.

9.6.3.1 F-test

Consider two mathematical models A and B which are identical except that model B contains an additional parameter. Model A is said to be nested in model B. Put another way, model A is nested in model B, since the parameters of model A are a subset of the parameters in model B. An example of a nested model includes the Michaelis-Menten equation and the Hill equation (Equations 9.3 and 9.4).

$$v = \frac{V_{max}[S]}{K_m + [S]}$$

Equation 9.3   **Michaelis-Menten equation.**

$$v = \frac{V_{max}[S]^h}{K_m{}^h + [S]^h}$$

Equation 9.4   **Hill equation.**

The exponent, h, is termed the Hill coefficient and is an indication of the degree of cooperativity in the substrate dependence. If $h > 1$, there is positive cooperativity; if $h < 1$, there is negative cooperativity. By inspection, it is clear that when $h = 1$, the Hill equation reduces to the Michaelis-Menten model. That is, the Michaelis-Menten equation is a special case of the Hill equation. The Michaelis-Menten equation is nested in the Hill equation.

Another example is the equation for simple enzyme inhibition and the equation containing a non-inhibitable background (Equations 9.5 and 9.4).

$$v = \frac{v_0}{1 + \frac{[I]}{K_i'}}$$

Equation 9.5   **Simple 2-parameter enzyme inhibition equation.**

$$v = \frac{v_0}{1 + \frac{[I]}{K_i'}} + bg$$

Equation 9.6   **Simple 2-parameter enzyme inhibition equation plus an additional background term.**

In this instance, the former equation is obtained by setting the background (bg) parameter equal to zero.

As all experimental data contain a degree of inaccuracy, inclusion of an additional parameter will always improve the quality of the fit. This is because it provides additional capability to allow for the experimental error. However, if this is the only benefit that is conferred by the inclusion of the additional parameter, then model B does not represent a real improvement over model A. The additional parameter must represent a physical property that is not present in model A, for model B to provide a genuine improvement.

Where two nested models are fitted to a single set of data, an F-test can be used to test the null hypothesis that the additional parameter in the complex model does not provide a genuine improvement. Note that the value of F must take into account that additional parameters will always improve the quality. The value of F is calculated by measuring the change in residuals associated with moving from the simple model to the complex model, relative to the experimental error associated with fitting the complex model. Equation for the value of F is shown in Equation 9.7.

Note that the numerator is a measure of the change in moving from the simple model to the complex one, with $(p_c - p_s)$ degrees of freedom. The denominator reflects the smaller residuals associated with the fit of the complex model and has $(N - p_c)$ degrees of freedom. The F-test does not require replicates and can be used to select the simplest equation which accurately describes the experimental data and so yields the most appropriate from the nested models.

$$F = \frac{\frac{(R_s - R_c)}{(p_c - p_s)}}{\frac{R_c}{(N - p_c)}}$$

Equation 9.7   Expression for the F statistic.
Where R_s = residual sum of squares for the simple model, R_c = residual sum of squares for the complex model, p_s is the number of parameters in the simple model, p_c is the number of parameters in the complex model and N is the number of data points.

The value of F is then used to compute a probability (p-value) for the validity of the null hypothesis. If the probability is low enough, the null hypothesis is rejected. Where $p \leq 0.01$, then the use of the complex model is justified. If, however, $p > 0.5$, then the probability of the improvement in fit being due to chance alone is >5% and so the data set does not justify acceptance of the more complex model. When $0.05 \geq p > 0.01$, there is a probability that the improvement is due to chance alone of 1–5%. This result may be viewed as an equivocal result and so may warrant the collection of a larger data set, with repeat of the analysis.

9.6.3.2   Akaike Information Criterion

The Akaike information criterion (AIC) involves the concepts of information theory, maximum likelihood theory, and entropy. The equation for AIC is shown below (Equation 9.8).

$$AIC = 2p + N \cdot \ln\left(\frac{R}{N}\right)$$

Equation 9.8   Expression for the Akaike information criterion (AIC).
Where N is the number of data points, R is the residual sum of squares and p is the number of parameters.

However, the AIC may be inaccurate with a small sample size, particularly when $N/p < 40$. In this situation, a corrected AIC, termed AIC_c should be used (Equation 9.9).

$$AIC_c = 2p + n \cdot \ln\left(\frac{R}{N}\right) + \frac{2p(p+1)}{N-p-1}$$

Equation 9.9 Expression for the corrected Akaike information criterion (AIC$_c$).

From these relationships, we can observe that the larger the value of R, the higher the AIC will be. It is also clear that AIC increases with the number of additional parameters added to the model. Thus, lower values of AIC indicate a better fit. If a parameter is added to the model, the value of R will go down. If the parameter has a significant effect on the quality of fit, the increase in the first term (for the simple AIC expression) will be more than offset by a decrease in the second term.

The AIC can be used to compare any two models fitted to the same dataset. The models do not need to be nested, which means that the AIC is a very powerful method for comparing unrelated models.

9.6.3.3 Absolute Goodness of Fit

The methods above enable comparison of different models and although it is possible that one model may represent a significant improvement over the other, this does not necessarily mean that the better model truly gives an accurate description of the data. It is therefore important to understand the accuracy of the model, to obtain an absolute evaluation of the goodness of fit. The procedure involves comparing the lack of fit with pure error and can be applied to nonlinear regression analysis [8].

The calculated value of R is comprised of two elements, which are due to pure error and lack of fit (Equation 9.10).

$$R = R_{pe} + R_{lf}$$

Equation 9.10 Expression for the residual sum of squares combining pure error and error due to lack of fit.

The null hypothesis is that the observed lack of fit is obtained by chance alone. The value of R is simply the sum of the squares of the residuals, and the deviation of the replicates is taken as an estimate for R_{pe}: the mean of the replicates is calculated and R_{pe} is then calculated as the sum of the squares of the deviations from the means. R_{lf} is then calculated as the difference $(R - R_{pe})$. The value of F is then obtained from the contribution due to the lack of fit relative to that due to pure error (Equation 9.11).

$$F = \frac{\frac{R_{lf}}{(N_x - p)}}{\frac{R_{pe}}{N_x(r-1)}}$$

Equation 9.11 Expression for the F statistic for assessing absolute goodness of fit.
Where N_x is the number of different values of the independent variable and r is the number of replicates at each value of the independent variable. If there is unequal replication, the denominator is replaced with $R_{pe}/\sum(r-1)$.

The value of p is then calculated from F. If $p > 0.05$, then the probability of obtaining the lack of fit by chance alone is $>5\%$, and the model is retained. When $0.05 \geq p > 0.01$, there is a probability of 1–5% of obtaining this quality of fit. This may justify the rejection of the model, although as described above, a larger data set may be required to perform further analysis. Where $p \leq 0.01$, the model is not valid for the data set.

9.7 Minimum Significant Ratio

The minimum significant ratio (MSR) is a useful statistical parameter for assessing the reproducibility of potency estimates generated from concentration-response analysis. The MSR is defined as the smallest ratio between the potencies of two compounds that is statistically significant and so may be used to assess whether the potencies of several different compounds are truly different. The logarithmic values of these measurements are usually normally distributed and, $\log IC_{50}$ values are therefore often calculated and used for statistical analysis, including determination of the MSR (Equation 9.12).

$$MSR = 10^{2\sqrt{2s}}$$

Equation 9.12 Expression for the minimum significant ratio (MSR).
Where s is an estimate of the standard deviation of the log potency for a single compound.

The variability of s may be estimated in several different ways depending on the data available. For example, it is possible to assess either intra-run (within the run) or inter-run (between runs) variability. To estimate the intra-run variability, two repeats each containing at least 20 compounds are undertaken. The standard deviation, s, is estimated from the paired differences in the logarithmic potency across the two runs. The MSR should be <3 for an assay to be considered suitable and suggests that 3-fold differences between potencies are significant.

Inter-run variability can be calculated from at least six runs, where s is estimated from the logarithmic potency values across runs. This measure of MSR is more representative of the reproducibility of an assay since the inter-run variability is larger than the intra-run variability. Values for MSR in this case of <4 are often representative of a reproducible assay with 4-fold differences in potency being significant. A small number of outliers can increase the variability estimates described above and may elevate the MSR.

If we consider estimating assay variability from the standard deviation of log potencies from repeat experiments, where x_i represents repeat IC_{50}s for a control compound, then s can be determined by Equation 9.13 and substituted into Equation 9.12 to determine the MSR.

$$s = \sqrt{\frac{\sum (x_i - \overline{x_i})^2}{n - 1}} \tag{9.13}$$

Equation 9.13 Expression for the standard deviation of log potency from *n* repeats.

Using the median absolute difference (MAD) provides a robust estimate of variability which is less sensitive to outliers (Equation 9.14) and the equation for MSR can be modified to incorporate the MAD (Equation 9.15).

$$MAD = median(|x_i - median(x_i)|)$$

Equation 9.14 Expression for the median absolute difference (MAD).

$$\mathrm{MSR} = 10^{\left(2\sqrt{2}\left(\frac{\mathrm{MAD}}{0.6745}\right)\right)}$$

Equation 9.15 Expression for the robust MSR.

The MAD makes use of the median rather than the mean, which is less sensitive to outliers, and uses the absolute value instead of the square. These differences make the MAD more robust to outliers compared to the standard deviation. The 0.6745 scale factor is needed for the MAD to be a consistent estimator of the standard deviation under the normal distribution.

9.8 Two Independent Variables

Below is an example for the analysis of enzyme kinetic data utilizing nonlinear regression where there is more than one independent variable in a single data set. Consider that each of the mechanisms competitive, uncompetitive, and pure noncompetitive inhibition (Chapter 8) is nested into the model for mixed noncompetitive inhibition. Each of these simpler models may be compared in turn with the more complex model using the F-test, or each model can be compared using the AIC. There are two possible outcomes: one of the simple models gives a goodness of fit comparable to that for the complex model, this simple model best describes the mechanism of inhibition or none of the simple models gives a goodness of fit comparable to that for the complex model. In this case, the complex, mixed noncompetitive model best describes the mechanism of inhibition.

Additional to the tests described above, the following method may be used to confirm the mechanism. If the approach above indicated that the mechanism is mixed noncompetitive, then the rate equation is shown in Equation 9.16.

$$v_i = \frac{v_0}{\left(1 + \left(\frac{[I](K_{is}[S] + K_{ii}K_m)}{K_{is}K_{ii}(S + K_m)}\right)\right)}$$

Equation 9.16 Expression for the rate of reaction in the presence of a mixed noncompetitive inhibitor.
Note to simplify the expression in Equation 9.17, the terms K_{is} and K_{ii} have been used to signify the competitive and uncompetitive components, respectively.

The dissociation constants K_{is} and K_{ii} can be replaced with the reciprocals, the association constants, Z_{is} and Z_{ii}, to give Equation 9.17.

$$v_i = \frac{v_0}{\left(1 + \left(\frac{\frac{[I][S]}{Z_{is}} + \frac{K_m}{Z_{ii}}}{\frac{([S] + K_m)}{Z_{is}Z_{ii}}}\right)\right)}$$

Equation 9.17 Expression for the rate of reaction in the presence of a mixed noncompetitive inhibitor, replacing dissociation constants with the reciprocal association constants.

The data can then be analyzed by fitting the modified equation, and a standard Student's t-test, using the t statistic from Equation 9.18, to assess if the inclusion of a specific term is justified in Equation 9.17.

$$t = \frac{Z}{\mathrm{SE}_z}$$

Equation 9.18 Expression for the t statistic for Student's t-test.
Where SE_z is the standard error of Z.

The value of t is then used to calculate the probability, p, that $Z = 0$. The number of degrees of freedom is $(N - p)$. More sophisticated data analysis methods can also be applied to data sets, which may contain fewer numbers of data points, to extract parameter values that are as, or more reliable, than those estimated from routine approaches. Two approaches are described below: global fitting (Section 9.8.1) and reducing and repeating (Section 9.8.2).

9.8.1 Global Fitting

Global fitting can be used to determine parameter values for a family of related curves when these curves all share at least one common parameter. For example, a family of separate concentration-response curves for different inhibitors against the same enzyme in the same experiment would all be expected to share a common v_0 (uninhibited control rate). The global fitting returns one (global) best-fit value for each shared parameter and a separate (local) best-fit value for each non-shared parameter.

Global fitting allows the number of data points in each curve to be reduced, so that the number of data points remains larger than the number of parameters, which is a requirement for nonlinear regression analysis.

Figure 9.5 shows a global fit for 10 compounds in an enzyme assay, where the control rate is the shared parameter and the IC_{50} values are fitted locally. The IC_{50} values for each compound in the fit above, where only two concentrations are used, are all within 2-fold of the IC_{50}s, with overlapping 95% confidence intervals, for the corresponding fit using all 12 data points. This shows that equivalent information is obtained for good-quality data when global fitting is applied to reduced data sets. This allows the reduction in resources without necessarily compromising the precision.

9.8.2 Reducing and Repeating and Reducing Only

Reducing the number of data points and repeating the concentration-response attempts to improve efficiency and effectiveness for concentration-response assays within drug discovery. Broadly, the reducing and repeating strategy may be split into two types: reducing only and reducing and repeating.

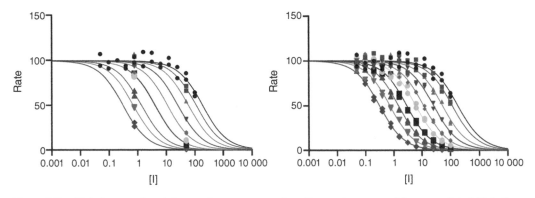

Figure 9.5 **Global to a reduced dataset (two concentrations for most compounds) versus global fit to the full dataset.**
Global fit for 10 compounds using 2 concentrations, selected to cover a wide range of potential IC_{50} values (left), and global fit for the same 10 compounds using the full 12 point concentration data. Similar values of IC_{50} for the 10 compounds are generated, demonstrating that when the data are of good quality, reduced data sets may be used to estimate parameter values more efficiently.

9.8.2.1 Reducing Only

This is the simplest, and perhaps most striking approach. It is simple, in that although the number of concentrations is reduced, the assay runs exactly as normal. It is striking that the number of data points can often be reduced by 50–75% without meaningfully affecting the estimated IC_{50}. This approach allows improved efficiency as project teams can make the same decisions more quickly and cheaply; or make better decisions by testing large numbers of compounds with the resource released by reducing or a combination of both. For example, a concentration-response curve could be reduced by 50% by removing duplicates. It also could be accomplished by selecting alternate concentrations from the initial full range. A combination of both approaches would give a reduction of 75%. Circumstances for which reducing is most attractive are when the emphasis is on screening to rank compounds, rather than on precision; there is little occasion-to-occasion variation in IC_{50} when a compound is repeated; on any single occasion, the concentration-response curve is very well-behaved.

Considering the high compound usage and throughput, the initial post-primary screening assays are prime candidates for reducing only. With half or less of the usual resource per compound, experience from many examples shows that interesting and inactive compounds are correctly classified.

The value of reducing only in such assays becomes higher if the proportion of inactive compounds is high. Reducing resources prevents resources from being wasted on compounds that will not progress.

As a result of using reduced concentration-response curves, it was recognized that in some (particularly well-behaved) assays, useful IC_{50} values could be estimated with as few as two concentrations. This has led to the development of the reduce and repeat approach.

9.8.2.2 Reducing and Repeating

The error between data sets is usually larger than the error within data sets, and so repeating an experiment is expected to improve the quality (precision) of the IC_{50}s generated in an experiment, as it incorporates a measure of the expected error. The reducing and repeating strategy, developed from earlier work on reduced data sets used in global fitting, improves the quality of measured IC_{50} values with less resource. The strategy is to replace 12-point concentration responses with 2-point concentration responses and then to repeat the measurement. This increases throughput by a factor of 3, gives better precision than the 12-point strategy, and reduces reagent consumption.

Repeating experiments is something that is common during drug discovery, and is done routinely in some assays, perhaps because the chemist has requested "an *n* of 2" (two IC_{50} determinations per compound). For other assays, a repeat might only be requested for interesting active compounds, or for compounds that were unexpectedly inactive when tested on the first occasion. The main value of having two IC_{50} values for a compound is that their average has less uncertainty than either of the two individual values. The uncertainty in the average is only 71% of the uncertainty in a single IC_{50} value.

This trend continues with increasing replicate numbers; the uncertainty in the average of three IC_{50} values is 58% of the uncertainty in a single IC_{50}; the uncertainty in the average of four IC_{50} values is 50% of the uncertainty in a single IC_{50}. Repeating improves precision, but by a diminishing degree as the number of replicates increases. However, the cost of repeating increases linearly.

Reducing and repeating, in some circumstances, will give more precise IC_{50} values using fewer data points in total than full concentration-response curves run on a single occasion. If the reduced protocol is a suitably small fraction of the full format, the additional costs encountered by repeating on another occasion can still be cheaper, as well as being the more precise strategy. Circumstances for which this reducing and repeating is most attractive are: the emphasis is more on precision than on screening; there is a lot of occasion-to-occasion variation in IC_{50} when a compound is repeated; on any given occasion the concentration-response curve is very well-behaved.

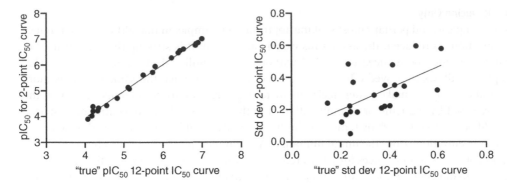

Figure 9.6 **Comparison of pIC$_{50}$ generated from 2-point and 12-point concentration-response fitting.**
There is good agreement between the pIC$_{50}$ values obtained from 2-point concentration-response fitting and those from 12-point concentration response fitting.

The graphs above (Figure 9.6) show for 20 cases of inhibition of kinases that there is a very strong correlation, with no evidence of systematic bias, between the IC$_{50}$ (from 12-point fitting) and the estimated value using 2-point reduced and repeated data. In 16 out of the 20 cases, the reducing and repeating strategy is more precise (lower standard deviation), despite the 3-fold increased throughput.

The reduce and repeat strategy should be used appropriately, to save resources where applicable, and to generate information on larger numbers of compounds. It is less advantageous for late-stage projects, where there is a much smaller set of compounds. Re-testing these often is routine, but as there are relatively few compounds, reducing the number of concentrations or replicates may be unnecessary with respect to cost and time. Additionally, at the later stages of drug discovery there often will be interest in the mechanism of action and hence in the detailed nature of the full curve shape, not just in the value of the IC$_{50}$. Here, reducing the number of concentrations would not only be unnecessary, but it would also be unhelpful.

References

1 Cleland WW. Statistical analysis of enzyme kinetic data. *Methods in Enzymology.* 1979;63:103–38.

2 Duggleby RG. A nonlinear regression program for small computers. *Analytical Biochemistry.* 1981;110(1):9–18.

3 Leatherbarrow RJ. *Enzfitter: A Non-linear Regression Data Analysis Program for the IBM PC and True Compatibles; Manuel:* Biosoft; 1987.

4 Storer AC, Darlison MG, Cornish-Bowden A. The nature of experimental error in enzyme kinetic measurements. *Biochemical Journal.* 1975;151(2):361–7.

5 Askelof P, Korsfeldt M, Mannervik B. Error structure of enzyme kinetic experiments: implications for weighting in regression analysis of experimental data. *European Journal of Biochemistry.* 1976;69(1):61–7.

6 Ellis KJ, Duggleby RG. What happens when data are fitted to the wrong equation? *Biochemical Journal.* 1978;171(3):513–7.

7 Lumley T, Diehr P, Emerson S, Chen L. The importance of the normality assumption in large public health data sets. *Annual Review of Public Health.* 2002;23:151–69.

8 McMinn CL, Ottaway JH. Studies on the mechanism and kinetics of the 2-oxoglutarate dehydrogenase system from pig heart. *Biochemical Journal.* 1977;161(3):569–81.

10

Molecular Interactions between Proteins and Small Molecules

10.1 Introduction

Enzyme kinetics can measure the strengths of interactions between proteins and small molecules and modification of the structure of the enzyme or the small molecule can divide the contribution of individual groups. This information may be valuable in designing new molecules with increased affinity for the target protein. However, measurements made using kinetics can be difficult to interpret for two main reasons: Firstly, the observed rate and equilibrium constants are rarely absolute values. A common example is the IC_{50} for an enzyme inhibitor, which often varies according to the identity and/or concentration of the substrate. Secondly, modification of the structure of a protein-ligand complex often involves changes in more than one feature of the interaction. For example, it is usually difficult to dissect the effects of the loss of a salt bridge (ionic interaction), from accompanying changes in solvation when removing one member of an ion pair by making a modification to the ligand.

The use of computational approaches has been advantageous in promoting our quantitative understanding of molecular recognition but may still lead to inaccuracies due to the large number of factors that must be accounted for. Stringent application of structure-activity relationships (SARs) employing isolated proteins can contribute to our understanding directly by elucidating binding constants, but also indirectly, by helping to improve the accuracy of computed values.

Below is a short review of the measured contributions to molecular recognition, with reference to the use of well-characterized systems that may facilitate interpretation. Usually, this requires knowledge of the 3D structure of the target protein and an ability to accurately measure absolute rate and binding constants.

Laboratory Guide to Enzymology, First Edition. Geoffrey A. Holdgate, Antonia Turberville, and Alice Lanne.
© 2024 John Wiley & Sons, Inc. Published 2024 by John Wiley & Sons, Inc.

10.2   Binding Affinity is a Function of Difference Energy

Consider the binding process between a protein and small molecule. Initially, these two partners are bound to water before they become bound to each other. Hence, the binding affinity is dependent upon the difference in energy between these two situations. So, the K_d value of the interaction reflects a number of factors:

1. The strength of hydrogen bonds in the complex compared to the combined strength of the hydrogen bonds for the individual partners to water.
2. The stability of salt bridges between the protein and ligand relative to the tendency for the individual ions to be solvated in water.
3. The van der Waals interactions compared with those in water.
4. The presence of hydrophobic interactions.
5. Unfavorable changes in entropy, which often manifest by the loss of translational and rotation freedom. The presence of favorable entropy changes may often reflect increased dynamics of other parts of the protein or the release of constrained water molecules.

These energy differences demonstrate the limited value of physical organic comparisons (the physicochemical properties defined by the structure of the molecule) for understanding contributions to binding affinity of small molecules in the absence of the protein target. However, utilizing enzyme kinetic approaches may allow the contribution of a substituent, S, to the binding free energy, ΔG_b, to be measured by comparing the dissociation constants of the parent compound A, and the substituent modified ligand A-S, as defined in Equations 10.1–10.3.

$$\Delta G_{b\,A-S} = R \cdot T \ln K_{A-S}$$

Equation 10.1   Expression of free energy of binding for the substituent modified ligand A-S.
R is the gas constant and T is the temperature.

$$\Delta G_{b\,A} = R \cdot T \ln K_{A}$$

Equation 10.2   Expression of free energy of binding for the parent compound A.

$$\Delta G_{b\,S} = \Delta G_{b\,A-S} - \Delta G_{b\,A} = R \cdot T \ln \left(\frac{K_{A-S}}{K_A} \right)$$

Equation 10.3   Expression to calculate the contribution to free energy of binding from substituent S.

Equation 10.3 provides an estimate of the incremental binding energy. Usually, this is a reasonable estimate when comparing inhibitors, but should be applied cautiously when employing substrates, as the binding energy may be employed to change the turnover number (k_{cat}), rather than the dissociation constant (K_s). For comparing modifications to substrates in this way, $\frac{K_m}{k_{cat}}$ for each substrate should be used instead.

It can be useful to translate the values of incremental binding energy into fold changes in binding affinity, as shown in Table 10.1, using Equation 10.4.

$$\text{Fold change} = e^{\left(\frac{\Delta G_b}{RT} \right)}$$

Equation 10.4   Expression of fold changes in binding affinity.

Table 10.1 Relationship between fold change in binding affinity and incremental binding energy. Values are calculated at 37 °C.

ΔG_b (kJ/mol)	ΔG_b (kcal/mol)	Fold improvement in binding affinity
1.80	0.43	2
4.15	0.99	5
5.94	1.42	10
11.88	2.84	100
17.81	4.26	1000
23.75	5.68	10 000

Of course, the incremental binding energy measured in this way is usually lower than the maximum binding energy, known as the intrinsic binding energy, as this maximum value can only be achieved if the group interacts optimally, without entropy loss compared to the parent compound.

The association of a molecule, A, with a protein can be described by a series of free energy changes, that may be described by Equation 10.5.

$$\Delta G_{b\,A} = \Delta G^i_A + \Delta G^e_A$$

Equation 10.5 **Expression for binding energy for ligand binding at a single site.**
$\Delta G_{b A}$ is the observed binding energy, ΔG^i_A is the intrinsic binding energy and ΔG^e_A is a correction term allowing for binding not being optimal.

The correction term ΔG^e_A comprises unfavorable contributions from:

1. The loss of entropy that results when two molecules form a complex.
2. Conformational changes that are required for complex formation.
3. Lack of complementarity between the protein and ligand (strain).

Conversely, the intrinsic binding energy arises from favorable contributions such as the formation of hydrogen bonds, ion-pair interactions, hydrophobic interactions, and release of bound water molecules.

Similarly, if a protein can bind ligands A and B at two different sites, the binding energy for the cross-linked molecule, A-B results in Equation 10.6.

$$\Delta G_{b\,A-B} = \Delta G^i_A + \Delta G^i_B + \Delta G_c$$

Equation 10.6 **Expression for binding energy for ligand binding at two different sites.**
ΔG^i_A is the intrinsic binding energy of molecule A. ΔG^i_B is the intrinsic binding energy of molecule B. ΔG_c is a correction term.

In this case, ΔG_c contains contributions from ΔG^e_A and ΔG^e_B, together with additional factors arising from the connection of A and B [1]. For example, the binding of B may decrease the freedom of the A moiety. Or, if B is not optimally aligned to the binding site, there may be a need to overcome additional strain, such that the binding of the B moiety is less favorable than binding of free B.

Use has been made of protein engineering approaches to estimate the strengths of different types of interaction [2]. Side chains on the protein are modified by site-directed mutagenesis and the apparent strength of the interaction is quantified by comparison of the measured kinetic parameters for the wild-type and mutant proteins. This approach, however, has some limitations, since removal of a side chain group may disrupt the hydrogen bonding network, have steric effects, perturb the local 3D structure, modify hydrophobic interactions, or change solvation. These effects often may be identified by inspection of the 3D structures for wild-type and mutant protein complexes with the ligand.

10.3 Interactions between Charged or Polar Groups

10.3.1 Electrostatic Interactions

Electrostatic interactions are difficult to quantify due to variation of the local dielectric constant, D, as well as the uncertainty around the contribution from polarization and desolvation. The value of D varies depending upon the local microenvironment. High values of D (around 60) are observed around charged groups on the surface of the protein, with low values (less than 10) in solvent-shielded hydrophobic regions [3].

Key Concept: Dielectric Constant

The dielectric constant defines the strength of the electrostatic interaction between two charges separated by a fixed distance. It is defined by the ratio of the dielectric permittivity of the substances, ε, to the dielectric permittivity of the medium that separates them, ε_0. Though water has a relatively high dielectric permittivity ($\varepsilon_{wat} = 78$ at 298 K), other components of the cell have permittivity's approaching the $\varepsilon = 2$ value of hydrocarbon crystal [4]. Initial estimates of the dielectric permittivity of a protein range from 2 to 80 [5–8].

The effect of varying the distance (r) between interacting groups is dependent upon their electrostatic properties (Table 10.2).

Hence, the apparent strength of a salt bridge may vary substantially, for a number of reasons, including:

1. The magnitude of the local dielectric constant [4–8].
2. Delocalization or concentration of charges on the protein side chain.
3. Release of solvating water upon complexation.
4. Interaction of charges with bridging water molecules or inorganic ions.
5. The precise position of the charges.

Table 10.2 The energy of different interacting groups.

Type of interacting group	Energy
Ions with net charges	$1/Dr$
Randomly oriented permanent dipoles	$1/Dr^3$
Ion interacting with an induced dipole	$1/Dr^4$
Permanent dipole interacting with an induced dipole	$1/Dr^6$

D is the dielectric constant and r is the distance between the interacting groups.

From site-directed mutagenesis studies, the deletion of a group involved in salt-bridge formation decreases binding affinity by 5–20-fold, when additional contributions from changes in solvation or changes in local 3D structure are ignored. This allows the prediction that the addition of a charged group into the precisely correct position on an inhibitor, in order to form a salt bridge with the protein, will likely increase binding affinity between 5- and 20-fold.

10.3.2 Hydrogen Bonds

Hydrogen bonds are directional electrostatic interactions that occur when a hydrogen atom, which is covalently bound to a donor, also forms a non-covalent interaction with an acceptor. The importance of hydrogen bonding in drug design has been recognized for many years [9–16], highlighting hydrogen bonding as a key determinant of binding affinity for drugs to their targets. However, the importance of individual hydrogen bonds cannot be determined from inspection of the structure of the protein-ligand complex. Proton donor and acceptor scales for groups often used in medicinal chemistry have been composed [13]. It is common for a single chemical group to interact with more than one donor or acceptor, or for groups to act as both donor and acceptor. This behavior allows the formation of hydrogen bond networks (Figure 10.1), which are observed within protein structures and also in ligand binding sites. It is also common for water molecules to form part of a hydrogen bond network, as the water molecule may form a bridge between the donor and acceptor.

Hydrogen bonds are critical for protein function, as they are strong enough to provide specificity, but weak enough to be broken and formed during catalysis. The relative weakness of hydrogen bonds in proteins may be partly due to the fact that the formation of a protein-ligand complex does not necessarily involve a net change in the number of hydrogen bonds. Consider a hydrogen bond donor on a free protein, (P-D-H), and an acceptor on a free ligand (L-A). In their free states, these groups would be interacting with solvating water. Upon complexation, there is an exchange of hydrogen bonding partners, but not net change in hydrogen bonds. Thus, the interaction between the protein and ligand may be driven predominantly by the change in entropy when bound water molecules are released into bulk solution

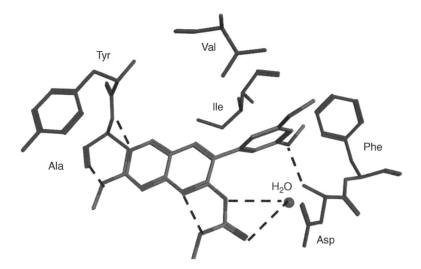

Figure 10.1 A hydrogen bonding network.
Hydrogen bonds (dashed lines) can be seen within the compound (orange), between the compound and protein side chains (purple), and between the compound and water molecules (magenta sphere).

$$P\text{—}D\text{—}H \cdots OH_2 + L\text{—}A \cdots H_2O \rightarrow P\text{—}D\text{—}H \cdots A\text{—}L + H_2O \cdots H_2O$$

Reaction scheme 10.1   Hydrogen bond formation between a free protein and ligand.
P-D-H is the hydrogen bond donor on a free protein. L-A is the hydrogen bond acceptor on a free ligand.
A dashed line represents a hydrogen bond.

Table 10.3   Example of hydrogen bond acceptors of increasing strength from weak to strong.

Table 10.3 (Continued)

The hydrogen bond acceptor is highlighted in red. X = O, S, NR.

(Reaction scheme 10.1). The release of several solvating water molecules and the exchange of weak hydrogen bonds for stronger ones will also favour the association of protein and ligand.

The contribution of a hydrogen bond to binding may be positive or negative depending upon the specific process and the solvent density [17]. For example, if two functional groups are completely removed from solvent during binding, then dissociation may be favored. However, partial removal from the aqueous environment may be optimal as this allows solvation of hydrogen bonding groups. Hydrogen bond strength is obviously affected by the identity of the interacting partners, and experiments in model systems have allowed ranking according to strength (Tables 10.3 and 10.4).

10.4 Nonpolar Interactions

10.4.1 van der Waals Interactions

It has been known for many years that the potential energy of interaction changes as a function of the separation distance, r. At short interatomic distances, there is a sharp increase in repulsion, approximating to an inverse twelfth power function (r^{-12} or $1/r^{12}$). There is also an attractive

Table 10.4 Example of hydrogen bond donors of increasing strength from weak to strong.

The hydrogen bond donor is highlighted in red. X = O, S, NR.

potential (van der Waals or London dispersion forces) which follows an inverse sixth power function (r^{-6} or $1/r^6$). These opposing forces combine to provide the overall potential energy, U, given by the Lennard-Jones potential (Equation 10.7).

$$U = \frac{A}{r^{12}} - \frac{B}{r^6}$$

Equation 10.7 Expression for the Lennard-Jones potential.
A and B are positive constants, such that $A = 4\varepsilon\sigma^{12}$ and $B = 4\varepsilon\sigma^6$, where ε is the depth of the potential well (usually referred to as "dispersion energy"), and σ is the distance at which the particle–particle potential energy, U, is zero (often referred to as "size of the particle").

Key Concept: Lennard-Jones Potential

The Lennard-Jones potential describes the potential energy of interaction between two particles as a function of the distance between them. It is a simple and accurate model for interatomic pair potentials which is often used in molecular dynamics simulations.

The $\frac{1}{r^6}$ term arises because the dispersion forces are due to the mutual induction of electrostatic dipoles. Thus, polarizability is an important factor in determining the strength of an interaction. The polarizability of common functional groups in proteins is found to increase in the order $H < O < OH < CH_2 < S < SH$.

The van der Waals distance is that providing the lowest potential energy, see Figure 10.2. The interaction energies for atoms and their van der Waals distances are given in Table 10.5.

Key Concept: van der Waals

van der Waals are weak distance-dependent interaction forces between molecules that arise due to the induction of electric dipoles. These occur when electron-rich regions of one molecule attract electron-poor regions of another, causing electric polarization. There are two forms of van der Waals: the weaker London Dispersion Forces (induced dipole–dipole) and the stronger Dipole–Dipole forces.

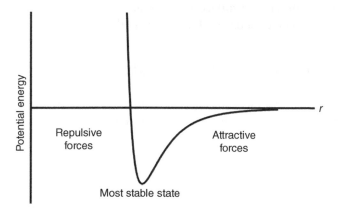

Figure 10.2 Potential energy diagram.
r is the separation distance. At a very close distance (r) the potential energy is very large and positive and so the interacting molecules repel each other. The potential energy is small and negative at a moderate distance and so the interacting molecules attract.

Table 10.5 Estimated interaction energies at the van Der Waals distances. Adapted from [18].

Interaction	ε_{ii} (kJ/mol)	ε_{ii} (kcal/mol)	r_{ii} (pm)
H···H (hydrocarbon)	0.079	0.019	281.52
O···O (carbonyl)	0.218	0.052	346.17
O'···O' (carboxyl)	0.611	0.146	366.65
N···N	0.573	0.137	408.66
C···C (carbonyl)	0.285	0.068	423.88
C···C (carboxyl)	1.891	0.452	363.43

ε_{ii} is the interaction energy for two identical atoms, r_{ii} is the van der Waals distance between two identical atoms.

10.4.2 Hydrophobic Interactions

Partitioning of solutes between aqueous and hydrophobic solvents is well-studied. The partitioning of a ligand between water and a hydrophobic site on a protein is controlled by three processes:

1. In bulk solution, water molecules become ordered around a hydrophobic solute in order to maximize the number of hydrogen bonds in the solvent. Driving a nonpolar molecule into a hydrophobic region of a protein allows an increase in entropy in the solvent. Additionally, there may also be an enthalpic component due to the formation of a hydrogen bonding network in bulk water, but not within the few water molecules fixed in ligand binding sites.
2. A smaller contribution is made by hydrogen bonding in the binding site, both to the protein and to trapped water molecules.
3. Dispersion forces lead to a slight increase in the strength of interaction because they are stronger in a hydrophobic environment, where there is a higher atom density, and the polarizability of the CH_2 group is greater than oxygen.

Hydrophobic interactions do not require complementary charged or polar groups and so contribute less to specificity than hydrogen bonds and ion pairs. The hydrophobic interaction strength is proportional to the magnitude of the contact surface area. 0.01 nm^2 of water-accessible surface area is estimated to produce a hydrophobic energy of 84–105 J/mol [19].

10.5 Changes in Hydrophobicity on Chemical Substitution

It was shown many years ago [20] that many substituents make a constant, additive contribution to hydrophobicity. The hydrophobicity constant, Π, for a group R is given by Equation 10.8.

$$\Pi = \log\left(\frac{P}{P_0}\right)$$

Equation 10.8 Expression for the hydrophobicity constant.
P_0 is the partition coefficient for the parent compound A and P is the partition coefficient for the derivative (A-R).

Partitioning experiments to measure the partition coefficients of small molecules are usually carried out using octanol as solvent. The experiments measure the partition of the compound between octanol and water (Equation 10.9).

$$Log\, P = Log\left(\frac{[\text{Drug in octanol}]}{[\text{Drug in water}]}\right)$$

Equation 10.9 Expression for Log P.

The value of log P can also be accurately estimated for many small molecules using computational approaches, where the values are calculated from fragment values, with the computed value usually designated cLog P.

Note that the value of log P varies according to pH if a group ionizes. This effect is taken into account in the distribution coefficient, D, which may be calculated (Equation 10.10).

$$Log\, D = Log\, P - Log(1 + 10^{(\text{pH}-\text{pK})})$$

Equation 10.10 Expression for Log D.

Or it is often measured as the distribution between octanol and buffer at pH 7.4 (Equation 10.11).

$$Log\, D_{7.4} = Log\left(\frac{[\text{Drug in octanol}]}{[\text{Drug in pH 7.4 buffer}]}\right)$$

Equation 10.11 Expression for Log D at pH 7.4.

It follows that $2.3RT\Pi$ is the incremental Gibbs free energy of transfer of the group –R from octanol into water, relative to the hydrogen atom (Table 10.6).

10.6 Entropy

When two molecules combine to form a complex, there is a loss in entropy, primarily because one partner effectively loses freedom of translation and whole-body rotation, but also some loss of entropy from both partners due to restriction of bond rotation and vibration. These unfavorable entropy changes may be offset to some extent by release of bound water molecules. A loss in entropy is relatively small when the size of a molecule is between 300 and 700 Da, so that adding substituents to a drug molecule generally has little effect on entropy changes unless there is a change in solvation.

If two molecules were to associate via a single complementary interaction, there would be a favorable enthalpic contribution, but unfavorable loss of entropy of translation and rotation. There would also be a gain in entropy as water of solvation is released upon complex formation. Conversely, when two molecules associate via multiple functional groups, they gain the energy of a large number of complementary interactions, but there is still loss of only one set of entropies. Many solvating water molecules may be released and so gain further entropy. Thus, although interactions between single functional groups are small, multiple interactions may be very strong, an effect known as the chelate effect.

The entropic contribution from the release of bound water is particularly important when molecules form a large number of weak interactions that are dispersed over a large area. Such interactions may occur in complexes between macromolecules, such as in protein–protein

Table 10.6 Values of Π and incremental free energy of transfer from octanol to water. Adapted from [20].

Group	Π	ΔG (kJ/mol)	ΔG (kcal/mol)	Expected fold change in affinity
$-CH_3$	0.56	3.31	0.79	3.6
$-CH_2CH_3$	1.02	6.02	1.44	10
$-(CH_2)_2CH_3$	1.5	8.91	2.13	32
$-(CH_2)_3CH_3$	2.0	11.88	2.84	100
$-(CH_2)_4CH_3$	2.5	14.85	3.55	320
$-CH(CH_3)_2$	1.3	7.7	1.84	20
$-CH(CH_3)CH_2CH_3$	1.8	10.67	2.55	62
$-CH_2Ph$	2.6	15.61	3.73	420
$-Cl$	0.71	4.18	1.00	5.1
$-Di-Cl$	1.42	8.37	2.00	26
$-Br$	0.86	5.06	1.21	7.2
$-CN$	-0.57	-3.35	-0.80	0.27
$-OH*$	-0.67	-3.93	-0.94	0.22
$-NHCOCH_3*$	-0.97	-5.73	-1.37	0.11
$-OCOCH_3*$	-0.01	-0.04	-0.01	0.98

Data are all relative to the H atom. * indicates bound to aliphatic hydrocarbons. Values of ΔG relate to 37 °C.

interactions. These systems are particularly challenging for drug design, where the aim is to find a small molecule with high affinity for only a small part of a large site.

Formation of a protein-ligand complex normally displaces water of solvation from functional groups on both partners into bulk water. This leads to an increase in entropy, which is favorable. Complex formation may also trap water of solvation, leading to an unfavorable decrease in entropy. Design of a ligand to increase the number of water molecules displaced should, therefore, lead to an increased binding affinity.

Methods to predict the magnitude of entropy changes assume that the bound molecule is highly constrained and then it is displaced into an environment where freedom is not restricted. However, this simplified situation is not representative of protein-ligand interactions in aqueous solution. Firstly, solvating water may retain some freedom when bound to functional groups on the free protein and free ligand. Secondly, a water molecule in the bulk solvent is constrained by hydrogen bonds. Estimates for the entropy change for the displacement of a water molecule have been made from that when ice melts into liquid water, which is around 22 J/K/mol [21]. This translates to 6.8 kJ/mol per molecule of water at 37 °C. Of course, the observed effect of displacing water will vary according to how constrained the water is when solvating functional groups, but each water molecule displaced upon complex formation is expected to increase binding affinity by a factor of around 14-fold.

References

1 Jencks WP. On the attribution and additivity of binding energies. *Proceedings of the National Academy of Sciences of the United States of America*. 1981;78(7):4046–50.

2 Johnson KA, Benkovic SJ. 4 Analysis of Protein Function by Mutagenesis. In: Sigman DS, Boyer PD, editors. *The Enzymes*. 19: Academic Press; 1990. p. 159–211.

3 Russell AJ, Thomas PG, Fersht AR. Electrostatic effects on modification of charged groups in the active site cleft of subtilisin by protein engineering. *Journal of Molecular Biology*. 1987;193(4):803–13.

4 Haynes WM. *CRC Handbook of Chemistry and Physics*: CRC Press; 2016.

5 Ramachandran GN, Sasisekharan V. Conformation of polypeptides and proteins. *Advances in Protein Chemistry*. 1968;23:283–438.

6 Gilson MK, Honig BH. The dielectric constant of a folded protein. *Biopolymers: Original Research on Biomolecules*. 1986;25(11):2097–119.

7 Nakamura H, Sakamoto T, Wada A. A theoretical study of the dielectric constant of protein. *Protein Engineering, Design and Selection*. 1988;2(3):177–83.

8 Svensson B, Jönsson B, Woodward C. Electrostatic contributions to the binding of Ca^{2+} in calbindin mutants: a Monte Carlo study. *Biophysical Chemistry*. 1990;38(1):179–83.

9 Wade RC, Goodford PJ. The role of hydrogen-bonds in drug binding. *Progress in Clinical and Biological Research*. 1989;289:433–44.

10 Jeffrey GA. *An Introduction to Hydrogen Bonding*: Oxford University Press; 1997.

11 Gilli G, Gilli P. *The Nature of the Hydrogen Bond: Outline of a Comprehensive Hydrogen Bond Theory*: OUP Oxford; 2009.

12 Taylor R, Kennard O. Hydrogen-bond geometry in organic crystals. *Accounts of Chemical Research*. 1984;17(9):320–6.

13 Abraham MH, Duce PP, Prior DV, Barratt DG, Morris JJ, Taylor PJ. Hydrogen bonding. Part 9. Solute proton donor and proton acceptor scales for use in drug design. *Journal of the Chemical Society, Perkin Transactions 2*. 1989(10):1355–75.

14 Abraham MH. Scales of solute hydrogen-bonding: their construction and application to physico-chemical and biochemical processes. *Chemical Society Reviews* 1993;22(2):73–83.

15 Laurence C, Berthelot M. Observations on the strength of hydrogen bonding. *Perspectives in Drug Discovery and Design*. 2000;18:39–60.

16 Laurence C, Brameld KA, Graton J, Le Questel J-Y, Renault E. The pKBHX database: toward a better understanding of hydrogen-bond basicity for medicinal chemists. *Journal of Medicinal Chemistry*. 2009;52(14):4073–86.

17 Ben-Naim A. The role of hydrogen bonds in protein folding and protein association. *The Journal of Physical Chemistry*. 1991;95(3):1437–44.

18 Warshel A, Levitt M. Theoretical studies of enzymic reactions: dielectric, electrostatic and steric stabilization of the carbonium ion in the reaction of lysozyme. *Journal of Molecular Biology*. 1976;103(2):227–49.

19 Fersht A. *Enzyme Structure and Mechanism*: W.H. Freeman; 1985.

20 Hansch C, Coats E. α-Chymotrypsin: a case study of substituent constants and regression analysis in enzymic structure–activity relationships. *Journal of Pharmaceutical Sciences*. 1970;59(6):731–43.

21 Atkins PW. *Physical Chemistry*: Oxford University Press; 1999.

11

Applications in Drug Discovery

11.1 Introduction

Enzymology has a critical role to play during early drug discovery (Figure 11.1). The application of a mechanistic understanding of enzyme systems is important in designing assays and establishing relevant screening conditions to enable the identification of suitable hit molecules. Post-hit identification, enzymology combined with biophysical measurements is crucial in understanding the mechanism of action of hits and leads. This powerful combination provides the chemist with the information required to design potent and efficacious compounds. Moving from *in vitro* thermodynamics and kinetics into the *in cellulo* setting is also important in order to better understand the physiological context of the inhibition or activation and to begin to fully exploit selectivity requirements. Enzyme kinetics also play a key role in the relationship between pharmacokinetics and pharmacodynamics (PK/PD). The utility of enzyme kinetics in each of these areas will be discussed below.

11.2 Pre-screening

Once a decision is made to focus on a particular enzyme as a drug target, considerations must be made when building the assay to ensure that it allows for identification of hits that modulate the enzyme's activity by the desired mechanism. A purely cellular environment for the enzyme and its substrates will maintain the relevant biological integrity but purified recombinant enzyme and model substrates that may not adequately resemble the native substrates may represent a more facile approach. Usually, drug-discovery campaigns include a combination of biochemical, biophysical and cell assays to identify and characterize modulators of enzyme targets and pathways of interest. To derive the most relevant insights from these approaches, with the aim of interpreting *in vitro* data in the disease setting, some key considerations are required.

Laboratory Guide to Enzymology, First Edition. Geoffrey A. Holdgate, Antonia Turberville, and Alice Lanne.
© 2024 John Wiley & Sons, Inc. Published 2024 by John Wiley & Sons, Inc.

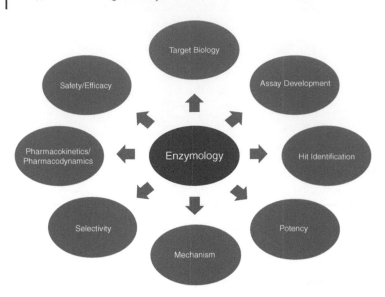

Figure 11.1 **The role of enzymology in early drug discovery.**

11.2.1 Enzyme Considerations

Firstly, a thorough understanding of the environment of the target of interest is necessary to enable the design of physiologically relevant assays. A decision is required about the use of recombinant enzyme domains versus systems likely to be of higher relevance and translatability, such as employing full-length enzymes and complex partners, which require a more detailed analysis of the underlying enzymology. The catalytic domains of full-length, multi-domain enzymes can sometimes be regulated by non-catalytic domains, which influence activity, substrate specificity, and inhibitor structure-activity relationships (SARs). Sometimes the use of the isolated domains can be poorly representative of the kinetic characteristics of the enzyme [1].

Key Example: Importance of Full-length Protein

In the complex physiological environment, full-length, multi-domain containing enzymes have catalytic domains that can be regulated by non-catalytic domains. These domains can control activity, substrate specificity, and inhibitor structure-activity relationships (SARs). A key example of such enzymes are the AKT serine/threonine protein kinases (AKT1–3). An exploitable site for inhibitor discovery was identified on the non-catalytic amino-terminal pleckstrin homology domain. A research strategy that utilized a recombinant protein encompassing only the catalytic domain of this membrane-associated protein kinase would have overlooked this important allosteric site [1].

Key Example: Effect of Complex Binding Partners

Protein binding partners can also influence the catalytic activity of the accompanying enzyme. A key example being the cyclin partners of the CDK enzymes, where activation of the cyclin-dependent kinases follows a two-step process comprising step 1: cyclin binding, followed by step 2: phosphorylation at a conserved threonine residue within the kinase activation loop. Such interactions and activation processes have the potential also to affect inhibitor binding and the associated structure activity relationships [2].

Detailed characterization and comparison of the mechanism and kinetic parameters between the different enzyme forms are required to be confident of the relevance of isolated domains with respect to the disease state of the enzyme. This is because domain architecture, post-translational modifications, activation state, and modifications such as affinity tags may all affect the kinetic behavior of the enzyme. Different binding partners may also influence the catalytic activity of the associated enzyme [2]. Carrying out assays in the presence of the native binding partner can improve the probability of identifying compounds that are more likely to be active in a physiological setting.

Additionally, multiple enzyme subunits may be involved in catalysis and so an understanding of the physiologically relevant protein complex is also key to ensuring that the correct form of the enzyme is present in the assay. Understanding whether specific complexes are associated with the disease-relevant state of the enzyme can help to drive selectivity.

This illustrates the importance of understanding the most physiologically relevant form of the enzyme to target. However, it is not always feasible to perform high throughput screening (HTS) with full-length protein and native substrates and so it is vital to explore the issues introduced using simpler systems and to try to correlate inhibitor activity to the cellular context. Another clear advantage of accumulating detailed knowledge of the molecular environment of the enzyme in the disease setting is the ability to prioritize specific mechanisms to target. For example, identifying inhibitors that are non-competitive with respect to the native substrate, where the cellular concentration of the substrate is high relative to K_m, may be beneficial. For instance, identification of kinase inhibitors that are either non-competitive or uncompetitive with respect to ATP, in order to avoid the challenge of high cellular concentrations of ATP, is a prudent approach. As mentioned in Chapter 8, enzyme inhibitors that function via an uncompetitive mechanism are ideal for inhibiting enzyme targets within signaling pathways because any accumulation of the substrate cannot reduce the degree of inhibition and so inhibitors following this mechanism may be less potent than competitive counterparts.

11.2.2 Substrate Considerations

Key Example: Importance of Physiological Substrates

There are many examples showing that interactions between macromolecular substrates and enzymes can take place at sites distal to the active site. These interactions can contribute significantly to the overall binding energy for the formation of the initial enzyme-substrate encounter complex, for example in the case of prothrombinase [3]. Replacement of the native substrate with a shorter peptide could affect both the enzyme's kinetic mechanism and the ability to identify inhibitors which might function by preventing formation of the ES complex by binding to these distal sites, particularly if specific conformational changes accompany the binding of the native substrate. Specific inhibitor pockets are, for example, revealed upon substrate binding in bacterial Glu-tRNA^{Gln} amidotransferases [4]. The importance of using native nucleosome substrates in preference over short, N-terminal histone tail peptides, has been recognized in order to identify cell-active small molecule inhibitors. For the histone methyltransferase, NSD2, the specificity of lysine methylation is directed by the choice of nucleosome substrate versus a histone octamer substrate in the presence of short ss-DNA or ds-DNA [5].

The nature and identity of enzyme substrates is another crucial consideration. There are many examples of binding interactions between macromolecular substrates and enzymes occurring at sites away from the active site, contributing to the binding energy for the formation of the initial enzyme-substrate complex. Replacing the native substrate with a shorter substrate might affect the kinetic mechanism and the ability to identify inhibitors that prevent formation of the ES complex by competing with binding to distal sites [3–5]. A detailed understanding of the catalytic mechanism and full consideration of the various enzyme-substrate interactions can afford control over the specific downstream pathways to be inhibited in a cellular context. This emphasizes an important observation that whilst many enzymes can catalyze turnover using surrogate substrates, the physiological relevance of this activity needs to be verified.

11.2.3 Enzyme Mechanism Considerations

Key Example: Understanding Mechanism to Inform Screening

A high throughput screening (HTS) campaign could be carried out twice: once in the absence of product, and subsequently in the presence of product or product analog. The first screen would reveal competitive or mixed-type inhibitors, whilst the second would identify uncompetitive inhibitors. For example, the rate-limiting steps of the bacterial enoyl reductases encoded by the *fabI* and *inhA* genes is the release of the product NAD^+. If the HTS campaign was carried out in the presence and absence of added NAD^+, or its analogue such as acetyl-pyridine adenine dinucleotide, hit compounds that bind more tightly to the enzyme-nicotinamide complex rather than the free enzyme might provide an alternative series of hit compounds to enter lead optimization [6].

Most enzyme-catalyzed reactions are rate-limited by either chemical or product release steps. Where the product release is rate limiting, the enzyme-product complex will be the principal enzyme form under steady-state conditions, especially at high substrate concentrations. In this case, a screening strategy where an enzyme-product complex is enriched by the assay conditions would be favored [6]. A HTS would probe a predominant enzyme species with a partially blocked active site, such that active compounds would inhibit the enzyme by preventing product release. An inhibitor of this type would generally produce uncompetitive inhibition of the enzyme. Crucially, screening a dominant enzyme-product complex samples different inhibitor chemical space than an assay where the free enzyme dominates, and the potential formation of an inhibitory ternary complex may provide higher selectivity than binding to the free enzyme.

Consideration of the rates of binding and dissociation, governed by the rate constants k_{on} and k_{off}, can also be important in designing inhibitors that exhibit kinetic selectivity. Kinetic selectivity, in which the drug-target complex has a longer half-life than off-target-drug complexes is important to achieve a safe and long-lasting effect. Slow-binding inhibition kinetics are a feature of many marketed drugs, and the concept of residence time has been an important consideration in the translation into cellular and *in vivo* contexts. Irreversible, covalent inhibition is also gaining attention, where tuning the balance between the initial molecular recognition event and the chemical reactivity can lead to the desired target selectivity without compromising safety and concern for off-target toxicity.

Hit-finding assays can be built in a number of ways to identify inhibitors with specific mechanisms [7]. For example, targeting substrate-uncompetitive mechanism may be achieved by ensuring the majority of the enzyme in the assay exists in the ES form. Assays may be balanced to ensure that the full breadth of modes of inhibition may potentially be captured. This may be achieved by ensuring that the concentration of free enzyme, E, is equivalent to the concentration of the ES complex, by arranging for conditions where the substrate concentration equates to its

Michaelis–Menten constant. Assays may also be configured to discriminate against unwanted mechanisms. For example, it is possible to bias away from competitive inhibition by arranging for an excess of substrate to be present in the assay. The ability to modify the concentration of the enzyme species is permitted by a detailed understanding of the kinetic mechanism relative to the physiological substrate conditions. This is particularly relevant for multi-substrate enzyme-catalyzed reactions, where a comprehensive understanding of the mechanism, and consideration of the relative concentrations of enzyme species during the reaction requires elucidation of the kinetic mechanism, the order of substrate addition, and the order of product release and any apparent substrate cooperativity. This is accomplished by undertaking product and dead-end inhibition studies, using kinetic isotope effects, and if appropriate, identifying covalent reaction intermediates. Understanding the molecular enzymology of selected targets enables the design of assays in a way that enables identification of mechanistically desired chemical equity that is targeted to physiologically relevant enzyme species.

Key Example: Importance of Detailed Mechanistic Studies

The dissection of the mechanism and understanding of the relative concentrations of the different enzyme species present at any point during the reaction requires the determination of the order of substrate addition and product release and understanding any apparent substrate co-operativity. This can be achieved through a combination of substrate concentration response experiments, product, and dead-end inhibition studies, kinetic isotope effects, and, where relevant, identification of covalent reaction intermediates. This comprehensive analysis was carried out for cathepsin C [7] and serves as a strong example for understanding of enzyme mechanism.

11.3 Post-screening

Following the successful discovery of hit molecules, their inhibitory effect must be measured to ensure that chemical optimization can be applied to those compounds with the highest probability for development. Traditionally, this has been accomplished by measuring the IC_{50}. Although these experiments are straightforward to perform, as emphasized throughout this book, it should be remembered that the IC_{50} itself may not be a true reflection of inhibitor affinity. The measured value may change significantly, varying with the experimental conditions, particularly the concentration of the substrate.

11.3.1 Measuring Potency

Potential differences may arise from changes in buffer components, such as pH and ionic strength. Additionally, binding is dependent upon protein flexibility, and changes in conformation may occur during induced fit, where the binding site of the enzyme adapts to the shape of the ligand. The magnitude of the IC_{50} may vary with substrate identity and enzyme concentration. Often, in conventional analyses of IC_{50} values, enzymes, substrates, and inhibitor are incubated for a fixed time period, meaning that any time-dependent inhibition effects are not detected in the assay. The situation inside a cell is different, as an enzyme target will experience regular exposure to its substrates but only temporary exposure to an added inhibitor, due to the compound's pharmacokinetic profile. A more useful approach to understand enzyme inhibition of hit or lead compounds may be to investigate any time dependency, for example, by pre-incubating the enzyme and inhibitor, followed by its dilution, before assessment of residual activity by added substrates. This would expose tight binding or time-dependent inhibition and may allow dissection of time-dependent inhibition versus a primary target, which may not occur with secondary targets.

Key Concept: The Importance of Protein Flexibility

Proteins are dynamic and there is constant exchange between different conformational states with similar energies. Binding sites are often characterized by regions of both high and low mobility. Binding exploits this protein flexibility and opportunity for dynamic rearrangement, where changes in conformation occur during induced fit – the influence of the ligand brings about a change in shape of the binding site to accommodate the ligand. This flexibility and changes associated with changes in substrate identity and concentration may affect the value of IC_{50}. This inherent flexibility can be exploited in drug design and plays an important role in the design of new molecules [8].

As mentioned in Section 11.2.3, screening assays sample a number of different enzyme species during the catalytic cycle, as substrates are bound, intermediates are formed, and products are released during the reaction. Inhibitors may potentially bind to a number of these enzyme forms. Enzymes may also be modified before or during the reaction, for example, different post-translational modifications, redox states, or partial degradation products may also be present [8]. These enzyme forms may have different catalytic efficiencies and sensitivities to test compounds. As a result, they may contribute to the measured IC_{50} in unclear ways. This serves to re-iterate the requirement for rigorous quality control assessments as detailed in Chapter 3, in order to understand the contribution of different enzyme forms to the observed enzyme activity and hence the degree of inhibition.

11.3.2 Reversibility

Enzyme inhibitors can act either reversibly or irreversibly, and these basic principles have been described in Chapter 6. It is clear that the time-dependent properties of enzyme inhibitors represent a continuum and compounds that covalently inactivate an enzyme may actually demonstrate greater reversibility than noncovalent compounds that can generate remarkably tight binary complexes. These types of inhibition show very different characteristics both *in vitro* and *in vivo*. Reversibility is not an absolute phenomenon, since inhibition considered to be irreversible may be reversible over a period of time that is much greater than the usual assay time. The operational definition that classifies inhibition as irreversible is the loss of enzyme activity caused by an inhibitor that is not restorable over the timescale of the enzyme activity assay.

Thus, the distinction between reversible and irreversible inhibitors is made on kinetic grounds and so it is crucial to recognize the importance of analyzing time courses for inhibitors suspected to be slow binding or irreversible. This highlights the requirement for kinetic analysis, rather than using single time points for concentration-response data, to avoid mis-annotation of potency and misleading information as to the mechanism of action (MoA). It is necessary to characterize the rate constants of the steps involved in forming the enzyme-inhibitor (EI) complexes. By following the progress curves for inhibition, in reactions starting with enzyme, it is possible to distinguish between the two different mechanisms of slow-binding. Another useful approach is to carry out jump-dilution experiments, where the pre-equilibrated EI complex is diluted rapidly into the assay, and recovery of enzyme activity is monitored as the inhibitor dissociates from the complex.

Irreversible, covalent compounds cannot be ranked in terms of efficacy using an equilibrium dissociation constant, K_d or K_i. This means that IC_{50} values are also not useful in ranking these compounds. Clearly, any irreversible inhibitor added at half the enzyme concentration will, as long as the time for reaction has been long enough, covalently modify all of the enzyme and so a concentration equal to half of the enzyme concentration will lead to a 50% decrease in rate. It is still possible to carry out concentration responses with irreversible compounds under fixed conditions of substrate concentration and time, although this is not recommended to compare irreversible inhibitor effectiveness. Instead, the kinetic parameter $\frac{k_{inact}}{K_I}$, determined from time-course experiments with varying inhibitor concentrations, is used. The advantage of using a kinetic approach to measure $\frac{k_{inact}}{K_I}$ is two-fold: it uses valid parameters to describe the inhibition, and it allows the efficacy of the irreversible inhibition to be dissected into the initial molecular recognition and subsequent chemical reaction steps. The ability to characterize irreversible inhibitors in this way is essential to allow modulation of the balance between intrinsic compound reactivity, which may lead to a lack of selectivity, and initial noncovalent interactions, which can help to improve recognition of the target protein. Thus, the availability of these parameters during lead optimization allows the chemist to tailor these requirements to mitigate concerns over reactive warheads and potential off-target toxicity [9, 10].

Key Example: Covalent Enzyme Inhibition

Covalent compound-enzyme interactions can provide significant advantages in terms of potency and sustained effect. However, off-target binding to related or even unrelated enzymes through reactive groups, usually having electrophilic properties, must be avoided. Hence, the selectivity of irreversible inhibitors is a crucial component during optimization. Utilizing glutathione reactivity and measuring $\frac{k_{inact}}{K_I}$ values is important in guiding selectivity development. The use of quantitative mass spectrometry to globally map targets and non-specific off target binding of covalent inhibitors in human cells is also invaluable [9, 10].

11.3.3 Inhibitor Mechanistic Characterization

Hit compounds may demonstrate a number of different mechanisms of inhibition. Competitive inhibition is common, as inhibitors are often either designed to mimic the substrates of an enzyme or to bind at the substrate-binding site. Other mechanisms are also possible, although uncompetitive inhibition is usually rarer and occurs when the inhibitor binds only after the substrate has bound to the enzyme.

Once inhibitors have been identified during primary screening, it is important to understand how the concentration of substrates affects the measured IC_{50} values to determine the mechanism of inhibition. This can then be used to extrapolate the observed activity in isolated protein assays to cell assays, where the concentration of substrates may be different. It also allows the assessment of selectivity, as the magnitude and direction of inhibitor selectivity can vary according to the mechanism of inhibition and the substrate concentration at which selectivity is assessed.

Initial experiments to determine the mechanism of inhibition are often undertaken using two substrate concentrations, above and below the K_m of the varied substrate. While this may give an indication of the mechanism of inhibition, it is recommended that full matrix-based experiments covering a range of substrate and inhibitor concentrations are undertaken, so that reliable K_i values can be estimated.

11.3.3.1 Combining Enzyme Kinetics and Biophysics

Combining or supplementing enzyme kinetic studies with biophysical methods can add value at many stages in the drug discovery process. Biophysical methods have been increasingly used in this way to enhance the understanding of the molecular MoA of test compounds. This combination extends through initial reagent characterization for protein and tool compound evaluation, assay development, and choice of HTS conditions to more detailed MoA studies during lead optimization. These highly informative studies early in each of these phases have the advantage of increased impact beyond that delivered by each alone and potentially allow for cost savings to be realized due to unsuitable reagents, incorrect assumptions about tool compounds, and inappropriate assay configuration to detect desired MoAs. A combination of biophysical methods to interrogate binding kinetics and protein integrity, coupled with detailed dissection of the enzyme catalytic mechanism and effects of intervention by small-molecule modulators, provides a comprehensive amount of detail at the molecular level [11].

For many soluble target classes, biophysical measurements are relatively straightforward, as the preceding steps of protein expression and purification allow the generation of large amounts of high-quality protein. This enables the combination of target-specific enzyme assays and biophysical methods that can answer specific questions and measure specific parameters. This information can be collated at several stages of drug discovery enabling assay development, understanding the output from primary screening, and establishing the key criteria on which to focus secondary screening and detailed characterization. Biophysical assays often can be developed on timelines that match these other screening activities, and so facilitate integration of the two disciplines for increased value during hit identification. A major advantage of the combination of these methods is the ability to identify true hits rapidly and reliably, reducing resources potentially wasted by following undesired mechanisms, or false positives [12].

Key Example: Combining Enzyme Kinetics and Biophysics

The combination of enzyme kinetic data with biophysical measurements is incredibly powerful in understanding the mechanism of action of compounds. For example, ITC experiments can very quickly provide information on the affinity of compounds binding to different enzyme forms, helping to assign mechanism of inhibition. In the left-hand panel, a test compound was titrated into different cofactor bound forms of EACPR, showing that the tightest binding was observed for the NADH bound form. In the right-hand panel SPR data was used to undertake transition state analysis demonstrating that ponatinib binding to FGFR1 is associated with a much larger unfavorable transition state enthalpy, and a 2.3 kcal/mol larger transition state free energy barrier compared to the Parke-Davis compound (PD173074) labeled PDA. This is in good agreement with the observed 70-fold slower binding kinetics for ponatinib [11].

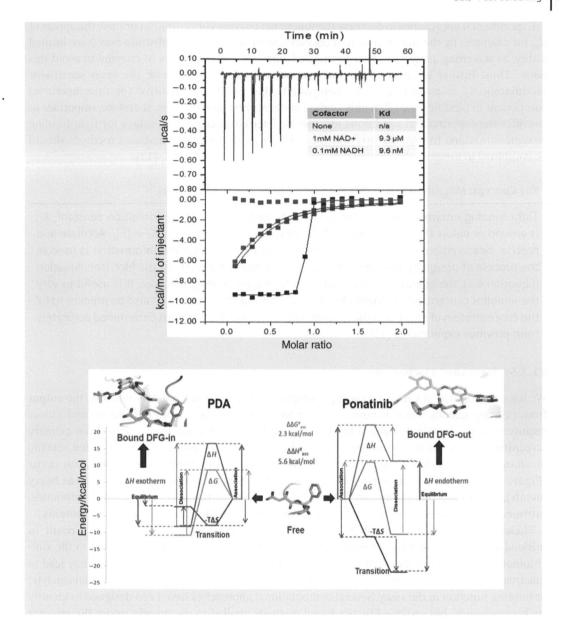

11.3.4 Tight Binding

The drug-discovery process often results in the identification of high-affinity molecules. High-affinity compounds are often difficult to characterize and rank correctly. It is important that true K_i values be compared, as IC_{50} values do not give an accurate representation of affinity under certain conditions, for example, situations of high-affinity binding when ligand depletion occurs. When this situation occurs, additional approaches must be taken to fully understand the true affinity.

Although the data-fitting approaches (described in Chapter 9) that do not assume that the free concentration of inhibitor equals the total, functional enzyme concentration, may be used, even these methods become unreliable as the affinity increases beyond around 10-fold below the functional enzyme concentration.

Experimental approaches to decrease the functional enzyme concentration or raise the apparent K_i, for example, by the use of increased concentrations of competing substrate may have limited utility, as screening assays are often configured with low concentrations of enzyme to avoid this issue. Thus, further lowering of the enzyme concentration can decrease the assay sensitivity. Modulating K_i' is possible only for compounds that have a competitive (or uncompetitive) component to binding, and often only limited shifts may be possible. It is, therefore, important to consider the experimental conditions in order to determine accurate K_i values for tight-binding enzyme inhibitors. In practice, this may mean biophysical and/or kinetic-based methods should be combined to measure the dissociation constant (K_d) and relate it to K_i [13].

Key Concept: Measuring Accurate K_i Values for Tight-binding Inhibitors

Tight-binding enzyme inhibitors are those compounds for which the inhibition constant, K_i, is around or below the concentration of the enzyme used in the assay $K_i \ll [E]_t$. Accurate and precise measurement of K_i values is particularly important when the information is used in the process of designing the next inhibitor. It is valuable to directly fit the Morrison equation (Equation 6.7), the quadratic expression describing tight-binding behavior. It is useful to vary the inhibitor concentrations around the enzyme concentration and it can also be prudent to fix the concentration of enzyme in the determination of K_i once it has been determined accurately from previous experiments [13].

11.3.5 Undesired Mechanisms

We have highlighted in Chapter 6 the types of spurious mechanisms that may infiltrate the output from primary screens. There is now a clear understanding of these types of behavior and a clear requirement to identify this and remove compounds showing undesired activity from primary screening hit lists. As has been described, and perhaps unsurprisingly, there are many specific mechanisms by which so-called nonspecific inhibition, or promiscuous inhibition, can occur (Figure 11.2). These mechanisms include common technology hitters, impurities (such as heavy metals), compounds that precipitate, compounds that cause aggregation, redox-active compounds, intrinsically reactive compounds, and compounds that bind preferentially to unfolded proteins.

These compounds, termed pan-assay interference compounds (PAINS) [14] may result in misleading IC_{50} values. This is because the inhibitory activity is not directly related to the stoichiometric inhibition of the protein, but often results from secondary effects. These may lead to inactivation, denaturation, or removal of active protein from solution or prevent its usual catalytic or binding function in the assay. Several computational approaches have been designed to identify such compounds, but enzyme kinetics-based methods applied to the protein under the relevant assay conditions are the preferred method for annotating such compounds.

There are several warning signs and key checks that can easily be carried out to help to identify PAINS. These include assessing the steepness of the slope of the inhibition concentration-response curve, investigating the time dependency of the inhibition, and checking for enzyme concentration-dependent changes in IC_{50}. Monitoring the effect of increasing detergent concentration and assessing the potential for inhibition in other assays with unrelated enzymes can be valuable. Characterizing compound reactivity with reducing agents and monitoring the effects on protein stability and unfolding, combining enzyme kinetics with biophysical methods is also extremely powerful.

It is essential to position these assays soon after initial hits have been detected, so that compounds potentially displaying these unwanted mechanisms can be assessed, annotated, and prioritized accordingly. These experiments are particularly useful when promiscuous activity occurs only

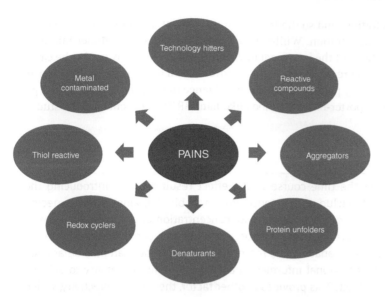

Figure 11.2 **Types of pan assay interference compounds that may inhibit enzyme reactions via undesired mechanisms.**

at concentrations much greater than those that are required for true, specific, binding-based inhibition. In this case, compounds may still be progressed through the drug-discovery process. The important consideration is that an understanding of the propensity for these types of liability is generated early, so that an informed decision-making process can occur, and compounds possessing mechanisms such as those listed above are not automatically ruled out from further optimization but considered on the balance of the activities observed.

11.4 *In Vitro* Versus *in Cellulo* Enzyme Kinetics

Conventional drug discovery typically involves initial chemical optimization driven by thermodynamic estimates of potency from *in vitro* assays. Ultimately, there is a requirement for a correlation between these *in vitro* measurements and cell-based estimates. There are many differences between a simple isolated protein system and the *in cellulo* setting and some of the key differences are highlighted in Table 11.1.

From these key differences, many of the assumptions usually valid in analyzing enzyme kinetic data do not hold in the cellular setting. For example, the substrate concentration may not be in large

Table 11.1 **Differences between *in vitro* and *in cellulo* enzyme kinetic measurements.**

In vitro	*In cellulo*
Homogenous system	Heterogeneous system
Often non-physiological enzyme construct	Relevant enzyme form
No crowding effects	Molecular crowding effects prevalent
No constraints to three-dimensional movement	Maybe constraints to molecular motion
$[E] \ll [S]$	$[E]$ may not always be $\ll [S]$
No turnover/resynthesis of enzyme	Turnover/resynthesis rates may be important
No clearance of compound	Compound may be cleared from target compartment

excess of the enzyme concentration, and so the free substrate concentration may not be assumed to equal the total substrate concentration. Whilst the *in vitro* analysis is done under initial rate conditions with the concentrations of the intermediates at steady state, this may not be the case in the cell-based setting. However, there is a strong drive to move beyond *in vitro* characterization and to begin to fully use cell-based assay parameters that can influence the chemical optimization process. Technologies, including reporter-based methods like nanoBRET, have enabled the study of cellular enzyme kinetics.

Key Concept: Importance of Kinetic Information for PK/PD Modelling

PK/PD models aim to predict the time course of the effect resulting from introducing the drug dose. Mechanistic models including the consideration of binding kinetics are expected to have some advantages over models in which only concentration and affinity are considered. This allows time-dependent changes in target occupancy to be incorporated. Additionally, key factors such as the rate of degradation and resynthesis of the target can also be applied to the models, which provide additional information on the PK profile necessary to achieve the desired pharmacological effect. This provides another factor, the kinetic selectivity, which should be considered when developing drugs for specific targets [15].

Kinetic information is vital for influencing PK/PD modeling in order to predict dosing conditions for candidate drugs. It must be remembered that the human body represents an open thermodynamic system and that cells adapt to changes in the environment rapidly. When a drug is administered to a patient, there are two important processes controlling the delivery of the compound to the site of action. Firstly, PK describes the way the body interacts with the compound to bring about its elimination from the systemic circulation. PD describes the effect the drug has on the body as a function of its concentration. Previously, the therapeutic effect has been modeled using equilibrium-based PK/PD models. However, it is also important to consider kinetic effects such as the time dependence of compound uptake and distribution, the rate of plasma-protein binding, the kinetics of target interaction and dissociation, as well as the flux of endogenous substrate and the rates of degradation and resynthesis of the enzyme [15].

A further consideration in applying enzyme kinetics to drug discovery is drug-target selectivity. Selectivity represents the interaction of the compound with the target protein compared to its binding to other, off-target, proteins in the system. Understanding selectivity is important to attempt to reduce toxicity that might arise from off-target interactions.

Traditionally, evaluation of target selectivity is usually undertaken by measuring the binding affinity of lead compounds against several potential targets under equilibrium conditions. Clearly, this does not take into account the kinetics of the various protein-ligand interactions. A combined strategy, including both equilibrium- and kinetic-based selectivity profiling, offers advantages during the early phases of the drug discovery process and should be considered when applying medicinal chemistry to design different or better leads. Usually, selectivity is calculated by taking the ratio of IC_{50} values, or more correctly K_i values, for the target enzyme versus off-target enzymes, which provides a measure of the thermodynamic selectivity. However, selectivity can also be predicted from kinetic measurements. This means that target occupancy and hence the on- or off-target biological effects can be different, depending upon the time of measurement. For example, consider the case where the K_i for a compound inhibiting the target is 10 nM, and for a related off-target enzyme is 100 nM. The thermodynamic selectivity would be 10-fold in favor of the target enzyme. However, if the rate constants for association and dissociation for the target and off-target enzyme were

1×10^5 M^{-1}s^{-1} and 0.001 s^{-1} and 1×10^7 M^{-1} s^{-1} and 1 s^{-1}, respectively, then the kinetic selectivity, in terms of residence time would be 1000-fold in favor of the target, and once filled, the half-life for dissociation from the target binding site would be 11.55 minutes, whereas for the off-target enzyme would be 0.69 seconds [16].

Key Example: Equilibrium and Kinetic Selectivity

This approach was applied to achieve selectivity profiling on the human adenosine receptors [16], where six AR antagonists were evaluated in a radiochemical displacement assay to determine their affinity and in a competition association assay to profile their binding kinetics versus three AR subtypes. This study demonstrated that two compounds (XAC and LUF5964) were kinetically more selective for the A$_1$R and A$_3$R subtypes, respectively, although they showed no selectivity in terms of their affinity.

This type of kinetic selectivity could arise from slow conformational changes and flexibility of protein and compound, or from additional interactions for the target enzyme. Adding the kinetic dimension can help guide decision-making during PK/PD studies. Target occupancy may be measured over time for both reversible and irreversible inhibitors and combined with studies on target turnover to allow an understanding of the duration of the biological effect. Clearly, the association and dissociation rates of the compound must be interpreted within the context of its PK and the degradation and resynthesis of fresh enzyme. The flux in the concentration of substrate and the mechanism of inhibition is also important, as association and dissociation rate constants may vary for different enzyme forms. Thus, robust modeling of PK/PD requires the consideration of enzyme kinetics, inhibitor binding kinetics, the PK of the small molecule, understanding of the flux in substrate concentration, the mode of inhibition, and an understanding of protein turnover. This will lead to greater understanding of drug efficacy and lead to safer and more effective medicines.

References

1 Lindsley CW, Zhao Z, Leister WH, Robinson RG, Barnett SF, Defeo-Jones D, et al. Allosteric Akt (PKB) inhibitors: discovery and SAR of isozyme selective inhibitors. *Bioorganic & Medicinal Chemistry Letters*. 2005;15(3):761–4.

2 Stevenson LM, Deal MS, Hagopian JC, Lew J. Activation mechanism of CDK2: role of cyclin binding versus phosphorylation. *Biochemistry*. 2002;41(26):8528–34.

3 Krishnaswamy S, Betz A. Exosites determine macromolecular substrate recognition by prothrombinase. *Biochemistry*. 1997;36(40):12080–6.

4 Harpel MR, Horiuchi KY, Luo Y, Shen L, Jiang W, Nelson DJ, et al. Mutagenesis and mechanism-based inhibition of *Streptococcus pyogenes* Glu-tRNAGln amidotransferase implicate a serine-based glutaminase site. *Biochemistry*. 2002;41(20):6398–407.

5 Li Y, Trojer P, Xu CF, Cheung P, Kuo A, Drury WJ, 3rd, et al. The target of the NSD family of histone lysine methyltransferases depends on the nature of the substrate. *Journal of Biological Chemistry*. 2009;284(49):34283–95.

6 Ward WHJ, Holdgate GA, Rowsell S, McLean EG, Pauptit RA, Clayton E, et al. Kinetic and structural characteristics of the inhibition of enoyl (acyl carrier protein) reductase by triclosan. *Biochemistry*. 1999;38(38):12514–25.

7 Schneck JL, Villa JP, McDevitt P, McQueney MS, Thrall SH, Meek TD. Chemical mechanism of a cysteine protease, cathepsin C, as revealed by integration of both steady-state and pre-steady-state solvent kinetic isotope effects. *Biochemistry.* 2008;47(33):8697–710.

8 Teague SJ. Implications of protein flexibility for drug discovery. *Nature Reviews Drug Discovery.* 2003;2(7):527–41.

9 Jöst C, Nitsche C, Scholz T, Roux L, Klein CD. Promiscuity and selectivity in covalent enzyme inhibition: a systematic study of electrophilic fragments. *Journal of Medicinal Chemistry.* 2014;57(18):7590–9.

10 Lanning BR, Whitby LR, Dix MM, Douhan J, Gilbert AM, Hett EC, et al. A road map to evaluate the proteome-wide selectivity of covalent kinase inhibitors. *Nature Chemical Biology.* 2014;10(9):760–7.

11 Klein T, Vajpai N, Phillips JJ, Davies G, Holdgate GA, Phillips C, et al. Structural and dynamic insights into the energetics of activation loop rearrangement in FGFR1 kinase. *Nature Communications.* 2015;6(1):7877.

12 Renaud J-P, Chung C-W, Danielson UH, Egner U, Hennig M, Hubbard RE, et al. Biophysics in drug discovery: impact, challenges, and opportunities. *Nature Reviews Drug Discovery.* 2016;15(10):679–98.

13 Murphy DJ. Determination of accurate K_I values for tight-binding enzyme inhibitors: an in silico study of experimental error and assay design. *Analytical Biochemistry.* 2004;327(1):61–7.

14 Baell JB, Nissink JWM. Seven year itch: pan-assay interference compounds (PAINS) in 2017—utility and limitations. *ACS Chemical Biology.* 2018;13(1):36–44.

15 Daryaee F, Tonge PJ. Pharmacokinetic-pharmacodynamic models that incorporate drug-target binding kinetics. *Current Opinion in Chemical Biology.* 2019;50:120–7.

16 Guo D, Dijksteel GS, van Duijl T, Heezen M, Heitman LH, AP IJ. Equilibrium and kinetic selectivity profiling on the human adenosine receptors. *Biochemical Pharmacology.* 2016;105:34–41.

Appendix A

Basic Math and Statistics

A.1 Algebra

Multiplication

$a \times b$ usually written as ab

$2a \times 3b = 2 \times a \times 3 \times b = 6ab$

$a \times a = a^2$

$a \times a^2 = a^3$

$2a^2 = 2 \times a^2 = a^2 + a^2$

Addition and subtraction

$3ab + 5ab = 8ab$

$6ab - 3ab = 3ab$

$2ab - 3c$ cannot be simplified further.

Laboratory Guide to Enzymology, First Edition. Geoffrey A. Holdgate, Antonia Turberville, and Alice Lanne.
© 2024 John Wiley & Sons, Inc. Published 2024 by John Wiley & Sons, Inc.

Expanding brackets

$$(2a + 6)(a - 2) = 2a^2 - 4a + 6a - 12 = 2a^2 + 2a - 12$$

A.2 Fractions

Simplification of fractions

$$\frac{(4a^2 - 4ab)}{(8ab - 8b^2)} = \frac{4a(a - b)}{8b(a - b)} = \frac{a}{2b}$$

Multiplication and division

$$\frac{a}{b} \times \frac{p}{q} = \frac{a \times p}{b \times q} = \frac{ap}{bq}$$

$$\frac{a}{q} \div \frac{b}{p} = \frac{a}{q} \times \frac{p}{b} = \frac{ap}{bq}$$

Addition and subtraction

$$\frac{2}{a} + \frac{3}{b} = \frac{2b}{ab} + \frac{3a}{ab} = \frac{(2b + 3a)}{ab}$$

A.3 Indices

Multiplication and division

$$a^p \times a^q = a^{p+q}$$

$$a^p \div a^q = a^{p-q}$$

$$(a^p)^q = a^{pq}$$

$$a^0 = 1, \quad \text{where } a \neq 0$$

$$(ab)^p = a^p b^p$$

Negative and fractional indices

$$a^{-p} = \frac{1}{a^p}$$

$$\sqrt[p]{a} = a^{\frac{1}{p}}$$

$$\sqrt[p]{a^q} = a^{\frac{q}{p}}$$

$$\sqrt[p]{\frac{a}{b}} = \frac{\sqrt[p]{a}}{\sqrt[p]{b}}$$

A.4 Logarithms

$$a^p = N \therefore p = \log_a N, \quad \text{where } p \neq 0 \text{ or } 1$$

Addition and subtraction

$$\log_a MN = \log_a M + \log_a N$$

$$\log_a \left(\frac{M}{N} \right) = \log_a M - \log_a N$$

Multiplication and division

$$\log_a M^p = p \log_a M$$

$$\log_a N = \frac{\log_b N}{\log_b a}$$

Natural logarithm (ln)

$$\log_e N \equiv \ln N \therefore \ln N = \frac{\log_{10} N}{\log_{10} e} = \frac{\log_{10} N}{0.434} = 2.303 \log_{10} N$$

A.5 Quadratic Equations

$ax^2 + bx + c = 0$ is the general form of a quadratic equation.

Factorising

$$x^2 - 5x + 6 = 0 \equiv (x - 3)(x - 2) = 0 \therefore x = 3 \text{ or } x = 2$$

Formula for solving quadratic equations:

$$ax^2 + bx + c = 0 \therefore x = \frac{\{-b \pm \sqrt{(b^2 - 4ac)}\}}{2a}$$

A.6 Straight Lines

$$y = mx + c$$

Represents a straight line with gradient m and intercept c on the y axis.

Since the values of m and/or c can be fractions, $ax + by + c = 0$ is also the equation of a straight line, although here c is not the intercept (since $y = -\frac{ax}{b} - \frac{c}{b}$).

A.7 Functions

A function is a rule which maps a single number to another single number.

A function of x is denoted $f(x)$. Example of the general form for different functions is shown below:

Quadratic function

$$f(x) = ax^2 + bx + c$$

Cubic function

$$f(x) = ax^3 + bx^2 + cx + d$$

Polynomial function

$$f(x) = a_n x^n + a_{n-1} x^{n-1} + a_{n-2} x^{n-2} + \ldots + a_2 x^2 + a_1 x + a_0$$

where $a_n, a_{n-1}, \ldots, a_1, a_0$ are constants, and n is a positive integer.

Exponential function

$$f(x) = a^x$$

where the variable is in the index.

A.8   Inequalities

The signs $<$ and $>$ signify inequalities, with $<$ indicating less than, such that $x < y$ means x is less than y and $>$ indicating greater than, such that $x > y$ means x is greater than y.

Consider the inequality

$$x > y$$

Adding or subtracting from both sides does not alter the inequality:

$$x + 3 > y + 3 \text{ and } x - 2 > y - 2$$

Multiplying or dividing both sides by a positive term does not alter the inequality:

$$3x > 3y \text{ and } \frac{x}{2} > \frac{y}{2}$$

Multiplying or dividing by a negative term reverses the inequality:

$$-3x < -3y$$

$$\frac{-x}{2} < \frac{-y}{2}$$

A.9   Differentiation

The process of finding the expression for the gradient of a straight line or curve.

When the variation of y is dependent on the variable x, then $\frac{dy}{dx}$ (first derivative) gives the rate at which y changes compared with x. $\frac{d^2y}{dx^2}$ is the second derivative, where $\frac{dy}{dx}$ is differentiated.

Differentiation is important in enzymology since the variation of [Product]([P]) is dependent on time and so the first differential of [P] with respect to time is the rate at which [P] changes with time, that is the velocity of the reaction.

If $y = f(x)$, the derivative of y or $f(x)$ is defined as

$$\frac{dy}{dx} = f'(x) \equiv \lim_{\delta x \to 0} \frac{\{f(x + \delta x) - f(x)\}}{\delta x}$$

$$\frac{d}{dx}(ax^n) = anx^{n-1}$$

$$\frac{d}{dx}(ax) = a$$

$$\frac{d}{dx}(a) = 0$$

Differentiating a function of a function

if $y = (ax + b)^n$

then $\dfrac{dy}{dx} = an(ax + b)^{n-1}$

Differentiating a product

$$\frac{d}{dx}(uv) = v\frac{du}{dx} + u\frac{dv}{dx}$$

Differentiating e^x

$y = e^x$

gives

$$\frac{dy}{dx} = e^x$$

e^x is the only function which is unchanged when differentiated.

Differentiating exponential and log functions (to the base e)

$$\frac{d}{dx}(e^{f(x)}) = f'(x)e^{f(x)}$$

For example, if

$y = e^{(2x+1)}$

then $\dfrac{dy}{dx} = 2e^{(2x+1)}$

$$\frac{d}{x}\{\ln f(x)\} = \frac{f'(x)}{f(x)}$$

For example, if

$y = \ln(3 + x^2)$

then

$$\frac{dy}{dx} = \frac{2x}{(3 + x^2)}$$

Differentiating an exponential function (when the base is not e):

$$\frac{d}{dx}(a^x) = a^x \ln a$$

The second derivative, obtained by differentiating $\frac{dy}{dx}$ or $f'(x)$ is denoted $\frac{d^2y}{dx^2}$

A.10 Integration

The process of finding a function from its derivative. The reverse of differentiation. Integration is important in enzymology since the integral of the rate of reaction with respect to time, between zero and time t gives the concentration of product formed over the period of time t.

$$\int \frac{d}{dx} f(x) \text{ then } dx = f(x) + c$$

where c is any constant

$$\int x^n dx = \frac{1}{(n+1)} x^{n+1} + c$$

Integration can be used to find the area under a curve, between two boundary points. $y = f(x)$ and the boundary points are $x = a$ and $x = b$, then area $= \int_a^b f(x) dx$

Integrating exponential functions (to the base e)

$$\int e^x dx = e^x + c$$

$$\int a e^x dx = a e^x + c$$

Integrating exponential functions (when the base is not e)

$$\int a^x dx = \frac{1}{\ln a} a^x + c$$

Note if a function is a derivative of a function, then the integral is known.

$$\frac{d}{dx} f(x) = f'(x) \therefore \int f'(x) dx = f(x) + c$$

For example, from above

$$\frac{d}{dx} \ln x = \frac{1}{x} \therefore \int \frac{1}{x} dx = \ln |x| + c$$

If the rate of decrease of x with time is proportional to x then $-\frac{dx}{dt} = kx$ integrating by separating the variables gives

$$\int \frac{1}{x} dx = \int -k dt \rightarrow \ln Ax = -kt$$

or

$$Ax = e^{-kt} \quad \text{or} \quad x = Be^{-kt}$$

where

$$B = \frac{1}{A}$$

The time taken for half of the original quantity to decay is called the half-life, $t_{\frac{1}{2}}$, and is given by:

$$\frac{1}{2} X_0 = X_0 e^{-kt} \rightarrow \frac{1}{2} = e^{-kt} \equiv 2 = e^{kt} \therefore t_{\frac{1}{2}} = \frac{\ln 2}{k} = \frac{0.6931}{k}$$

A.11 Series

An arithmetic series with n terms can be written as:

$$a, (a + b), (a + 2b), (a + 3b), \ldots, [a + (n - 1)b]$$

A geometric series with n terms can be written as:

$$a, ab, ab^2, ab^3, \ldots, ab^{n-1}$$

So, to calculate a dilution factor (b), for a series of n concentrations, where the first (a) and last concentrations are known.

$$b = \sqrt[n-1]{\left(\frac{\text{last concentration}}{\text{first concentration}}\right)} = \sqrt[n-1]{\left(\frac{ab^{n-1}}{a}\right)}$$

The use of this formula enables geometrically spaced concentrations to be used for studies involving enzymes, which often enables the range of useful concentrations to be covered most effectively.

The arithmetic mean of 2 numbers a and b is $\frac{1}{2}(a + b)$

The geometric mean of two numbers a and b is $\sqrt{(ab)}$. Often, it is more appropriate to use the geometric mean for parameters that must be positive, for example, dissociation constants and rate constants. This is because the measured value is likely to be correct within a certain multiple, not a certain number.

A.12 Statistics

Standard errors

Standard errors (SE) – are reported by most nonlinear regression software packages and are used to gain information about how precisely the best-fit parameter is known. These standard errors are asymptotic or approximate values and can be used to calculate approximate 95% confidence intervals (CI), which is a range of values that are predicted to contain the true value 95% of the time. Note in the context of nonlinear regression, the standard error is the same as a standard deviation.

$$95\%\text{CI} = \text{best fit value} - (t \times \text{SE}) \text{ to best fit value} + (t \times \text{SE})$$

where t is the value of the t statistic, found in tables (or the EXCEL function TINV (0.05, DF)), with the appropriate number of degrees of freedom (number of data points – number of parameters)

F-test

The F-test can be used to compare two "nested" models in order to help to reject a simple model in favor of a more complex one. Two models are nested if the simple model can be obtained from the complex model by fixing a parameter value in the complex model to zero or infinity, or by assuming that two or more parameter values are equal, e.g. $y = mx$ is nested into $y = mx + c$ by fixing c at zero.

The F ratio is calculated as follows:

$$F = \frac{\left\{ \frac{(SS_{simple} - SS_{complex})}{(P_{complex} - P_{simple})} \right\}}{\left\{ \frac{SS_{complex}}{(N - P_{complex})} \right\}}$$

where N is the number of data points, P is the number of fitted parameters, SS is the sum of squares reported by the regression program ($= c^2 \times$ number of degrees of freedom (DF $= N - P$)).

The F ratio can then be used to calculate a *p*-value (in this case the probability that the increase in the goodness of fit on moving from the simple to the complex model is by chance alone). This is done by looking at tables (see Appendix 3), or by using the EXCEL function FDIST (F, $DF_{numerator} = (P_{complex} - P_{simple})$, $DF_{denominator} = (N - P_{complex}))$.

A.13 Propagation of Errors

Propagation of errors must be accounted for when combining parameter values with associated standard errors:

Sums or differences: $X = A + B$ or $X = A - B$ then $\Delta X = \sqrt{(\Delta A)^2 + (\Delta B)^2}$

Products or ratios: $X = AB$ or $X = \frac{A}{B}$ then $\frac{\Delta X}{X} = \sqrt{\left(\frac{\Delta A}{A}\right)^2 + \left(\frac{\Delta B}{B}\right)^2}$

Functions: $X = f(X)$ or $X = X(A, \dots, B)$ then $\Delta X = \sqrt{\left(\frac{dX}{dA}\Delta A\right)^2 + \cdots + \left(\frac{dX}{dB}\Delta B\right)^2}$

A.14 Using Logged Values

Log-fitting of parameter values (that is fitting to the logarithm of the parameter) is often more appropriate for enzyme kinetic parameters such as K_m or K_d. This is because the distribution of the parameter values is not Gaussian (bell-shaped plot of frequency versus parameter value), but rather the distribution of the logarithm of the parameter is Gaussian. The approach also allows more meaningful estimation of confidence intervals, since with the log-fitting approach these cannot include negative values, which is more appropriate for parameters that cannot have negative values. It should be stated, however, that it may not be appropriate to use log-fitting for all parameters, since these parameters may already have a Gaussian distribution, without the transformation to log values. Testing the parameter distribution by simulation can help in the choice of data fitting procedure.

Log-fitting transformation for the Michaelis-Menten equation:

$$v = \frac{V_{max}[S]}{10^{(\log K_m)} + [S]}$$

The parameter value needs to be converted back from the logged value before it can be reported.

A.15 Precision, Accuracy, Significant Figures and Rounding

Precision and accuracy

Measurements always have some degree of uncertainty. Two terms often used to describe the uncertainty in measurements are precision and accuracy. These terms are often used interchangeably in everyday life; however, they have different meanings scientifically. Accuracy refers to the agreement between the measured value and the true value. Precision refers to the degree of agreement between several measurements of the same quantity and reflects the reproducibility of the measurement.

An example of precision and accuracy is given here for three thermometers, each taking five measurements of temperature of a solution at 37.0 °C.

1. Neither precise nor accurate: 34.6, 33.5, 33.2, 32.0, 38.2 °C
2. Precise but not accurate: 32.2, 32.1, 32.2, 32.1, 32.3 °C
3. Both precise and accurate: 37.0, 37.1, 37.1, 36.9, 37.0 °C

In another example, three people are playing darts. The aim is to hit the 20.

Antonia's darts (red) are neither precise nor accurate. Alice's darts (green) are precise, but not accurate. Geoff's darts (blue) are of course both precise and accurate.

Significant figures

Significant figures are used to indicate the probable uncertainty associated with a measurement. The significant figures in a number are all of the certain digits and the first uncertain digit.

Rules for counting significant figures:

1. Non-zero integers always count as significant figures.
2. Zeros
    (a) Leading zeros precede all of the non-zero digits, and do not count as significant figures. They indicate the position of the decimal point.
    (b) Captive zeros are zeros between non-zero digits and always count as significant figures.
    (c) Trailing zeros are zeros right at the end of a number and are significant only if the number contains a decimal point.
3. Exact numbers are numbers not obtained by measurement and can be considered as having an infinite number of significant figures.

Examples: 4.62 has three significant figures, 0.0032 has two significant figures, 1.005 has four significant figures, 200 has one significant figure and 2.00×10^2 has three significant figures.Rules for significant figures in mathematical operations:

1. Multiplication and division: the number of significant figures in the result is the same as the number in the least precise measurement used in the calculation.
2. Addition and subtraction: the result has the same number of decimal places as the least precise measurement used in the calculation.

Examples: $3.45 \times 1.6 = 5.52$, corrected to two significant figures is 5.5; $13.11 + 16.2 = 29.31$, corrected to one decimal place $= 29.3$.

Rounding

In many calculations, the final result will need to be rounded before the result is given to the correct number of significant figures.

Rounding: When rounding, the extra figures are carried through to the final result before rounding according to the following rules

1. If the digit to be removed is less than 5, the preceding digit remains the same.
2. If the digit to be removed is greater than or equal to 5, the preceding digit is increased by 1.
3. When rounding only the first number to the right of the last significant figure is considered.

Examples: 2.34 is rounded to 2.3, 1.46 is rounded to 1.5 and 5.349 is rounded to 5.3, all to two significant figures.

A.16 Dimensional Analysis

Mechanistic enzymology relies upon the use of equations derived from models of the physical behavior of enzymes in order to measure important parameters which define the systems under study.

Therefore, it is often very useful to be able to check the equations being used. By using a procedure known as dimensional analysis, it is possible to check that a derived equation may be correct. This powerful tool is often overlooked, but it can help to identify mistakes within equations and should always be used to check derived expressions. A simplification is to check using units for each parameter instead of the underlying dimensions.

There are five basic physical quantities used in biochemistry.

Physical quantity	Dimension	SI unit	Symbol
Length	L	meter	m
Mass	M	kilogram	kg
Time	T	second	s
Temperature	Q	Kelvin	K
Amount of substance	—	mole	mol

Note: The mole is strictly dimensionless but is often used for dimensional analysis purposes.

There are rules governing the use of dimensions.

1. Different dimensions cannot be equated, $L = M$ or m = kg $\times$
2. Different dimensions cannot be added or subtracted, $L + M$, or m + kg $\times$
3. Different dimensions cannot be greater than or less than each other, $L > M$, or m > kg $\times$
4. Different dimensions can be multiplied or divided L/T or m/s ✓
5. Dimensioned quantities cannot be exponents, 10^{2kg} $\times$
6. Logarithms of dimensions cannot be taken, $\log(10\,\text{s})$ $\times$

The usefulness of dimensional analysis is shown below for checking the Michaelis–Menten equation. We may think the equation is:

$$v = \frac{V_{\max}[S]}{K_m \cdot [S]}$$

We can check this by dimensional analysis, or by comparing units:

$$v = L^{-3}T^{-1}, \; \frac{V_{max}[S]}{K_m[S]} = \frac{L^{-3}T^{-1}L^{-3}}{L^{-3}L^{-3}} = T^{-1} \times$$

$$v = Ms^{-1}, \; \frac{V_{max}[S]}{K_m[S]} = \frac{Ms^{-1}M}{MM} = s^{-1} \times$$

We know that the dimensions do not equate so the equation must be wrong. The right-hand side must give the same dimensions as the left-hand side, and we can show that the correct equation satisfies this rule.

$$v = \frac{V_{max}[S]}{K_m + [S]}, \; L^{-3}T^{-1} = \frac{L^{-3}T^{-1}L^{-3}}{L^{-3} + L^{-3}}$$

$$Ms^{-1} = \frac{Ms^{-1}M}{M + M} \; \checkmark$$

Appendix B

Some Useful Key Formulas

CHAPTER MENU

B.1 Assay Reagent Calculations

Calculating concentrations

$$\text{Moles} = \frac{\text{mass}}{\text{molar mass}}$$

where Moles is the number of moles, mass is the mass of substance, and molar mass is the mass of 1 mol of substance.

$$M = \frac{\text{Moles}}{v}$$

where M is the molar concentration, Moles is the number of moles, and v is the volume in liters.

%v/v

$$\%v/v = 100 \left(\frac{\text{volume of sample}}{\text{total volume}} \right)$$

Example: 20 mL of DMSO added to 80 mL of water represents a 20% *v/v* solution of DMSO.

%w/v

$$\%w/v = 100 \left(\frac{\text{mass of sample}}{\text{total volume}} \right)$$

Example: 5 g of TCEP in a total volume of 100 mL represents a 5% *w/v* solution of TCEP.

%w/w

$$\%w/w = 100 \left(\frac{\text{mass of sample}}{\text{total mass}} \right)$$

Example: 10 g of salt in 90 g of water represents a 10% *w/w* solution of salt.

Laboratory Guide to Enzymology, First Edition. Geoffrey A. Holdgate, Antonia Turberville, and Alice Lanne.
© 2024 John Wiley & Sons, Inc. Published 2024 by John Wiley & Sons, Inc.

Dilution calculations

Calculations for the dilution of stock solutions can be carried out in two slightly different ways. The underlying mathematics involved is essentially the same, however, the methods may be taught differently, and their use based on preference. Some scientists will use the equation below (where c and v correspond to concentration, volume, and the subscripts i, and f relate to initial and final conditions) to calculate the final concentration or volume of the diluted solution.

$$c_i v_i = c_f v_f$$

Other scientists will calculate the dilution factor (DF) first before dividing the initial concentration by the dilution factor to obtain the final concentration. Or multiplying the initial volume by the DF to obtain to the final volume. The calculation for the DF is shown below and follows the same nomenclature as the equation above.

$$DF = \frac{v_f}{v_i} = \frac{c_i}{c_f}$$

Examples: the dilution factor for diluting 0.1 mL of compound in 9.9 mL of buffer is 100 (10/0.1); the volume of stock required to make a 1 in 500 dilution in a final volume of 1.5 L is 3 mL (1500/500).

Henderson–Hasselbalch equation

$$pH = pK_a + \log_{10}\frac{[A^-]}{[HA]}$$

where pH is the acidity of the solution (the negative logarithm of the $[H^+]$, pK_a is the negative logarithm of K_a, [HA] is the concentration of acid, $[A^-]$ is the concentration of conjugate base.

Beers Law

$$A = \varepsilon Cl = -\log_{10}T = \log_{10}\left(\frac{I_0}{I}\right)$$

where A is absorbance, ε is the molar extinction coefficient, C is the concentration, l is the pathlength, T is transmittance, I_0 is the incident light intensity, and I is the transmitted light intensity.

Correcting pD when replacing water with D_2O

$$pD = pH_a + 0.41$$

where pH_a is the reading on the normal pH meter.

B.2 Assay Statistics

Assay signal window

$$SW = \frac{(\overline{max} - 3SD_{max}) - (\overline{min} + 3SD_{min})}{SD_{max}}$$

where $\overline{max}$ and $\overline{min}$ are the average values for the maximum and minimum signals, respectively, SD_{max} and SD_{min} are the standard deviations of the maximum and minimum signals, respectively.

% CV

$$\%CV = \left(\frac{SD}{Mean}\right) \times 100$$

where SD is the population standard deviation and mean is the population mean.

Z′ factor

$$Z' = \frac{(\overline{\mathrm{max}} - 3\mathrm{SD}_{\mathrm{max}}) - (\overline{\mathrm{min}} + 3\mathrm{SD}_{\mathrm{min}})}{(\overline{\mathrm{max}} - \overline{\mathrm{min}})}$$

where $\overline{\mathrm{max}}$ and $\overline{\mathrm{min}}$ are the average values for the maximum and minimum signals, respectively, $\mathrm{SD}_{\mathrm{max}}$ and $\mathrm{SD}_{\mathrm{min}}$ are the standard deviations of the maximum and minimum signals, respectively.

% inhibition

$$\%I = 100 \left\{ \frac{(\mathrm{max} - X)}{(\mathrm{max} - \mathrm{min})} \right\}$$

where max is the maximum signal or rate, X is the measured signal or rate and min is the minimum signal or rate.

% control

$$\%C = 100 \left\{ 1 - \frac{(\mathrm{max} - X)}{(\mathrm{max} - \mathrm{min})} \right\}$$

where max is the maximum signal or rate, X is the measured signal or rate and min is the minimum signal or rate.

B.3 Curve Fitting and Calculation of Parameters

Exponential decrease

$$[\mathrm{A}] = [\mathrm{A}]_0 e^{-kt}$$

where $[\mathrm{A}]$ is the measured concentration at time t, $[\mathrm{A}]_0$ is the initial concentration, k is the rate constant, and t is the time.

Exponential increase

$$[\mathrm{A}] = [\mathrm{A}]_0 e^{kt}$$

where $[\mathrm{A}]$ is the measured concentration at time t, $[\mathrm{A}]_0$ is the initial concentration, k is the rate constant, and t is the time.

Simple 2-parameter inhibitor concentration response

$$v = \frac{v_0}{\left(1 + \frac{[\mathrm{I}]}{K_i'}\right)}$$

where v is the measured rate, v_0 is the control rate in the absence of inhibitor, $[\mathrm{I}]$ is the inhibitor concentration, and K_i' is the apparent inhibition constant.

Four parameter logistic equation

$$v_i = \frac{v_0}{1 + \left(\frac{[\mathrm{I}]}{\mathrm{IC}_{50}}\right)^h} + \mathrm{bg}$$

where v_i is the observed rate in the presence of inhibitor, v_0 is the control rate in the absence of inhibitor, IC_{50} is the inhibitor concentration giving 50% activity, h is the Hill slope, and bg is the uninhibited background rate.

Four parameter logistic equation fitting pIC₅₀

$$v_i = \frac{v_0}{1 + \left(\frac{[I]}{10^{pIC_{50}}}\right)^h} + bg$$

where v_i is the observed rate in the presence of inhibitor, v_0 is the control rate in the absence of inhibitor, pIC_{50} is the negative logarithm of the inhibitor concentration giving 50% activity, h is the Hill slope, and bg is the uninhibited background rate.

Half-life

$$t_{1/2} = \frac{\ln 2}{k_{obs}} = \frac{0.693}{k_{obs}}$$

where k_{obs} is the observed rate constant.

Residence time

$$\tau = \frac{1}{k_{off}}$$

where τ is the residence time and k_{off} is the dissociation rate constant.

Approach to equilibrium

$$k_{obs} = k_{on}[L] + k_{off}$$

where k_{on} is the association rate constant, k_{off} is the dissociation rate constant, and [L] is the ligand concentration.

Michaelis–Menten Equation

$$v = \frac{V_{max}[S]}{(K_m + [S])}$$

where v is the measured rate, V_{max} is the maximum rate, K_m is the Michaelis constant and [S] is the substrate concentration.

General expression for K_i' for a mixed-type inhibitor

$$K_i' = \frac{K_{is}K_{ii}(K_m + [S])}{K_m K_{ii} + [S]\,K_{is}}$$

where K_i' is the apparent inhibition constant, K_{is} and K_{ii} represent the competitive and uncompetitive inhibitor dissociation constants, respectively, K_m is the Michaelis–Menten constant and [S] is the substrate concentration.

B.4   Thermodynamics

Gibbs–Helmholtz equation

$$\Delta G = \Delta G^0 + RT\ln\left(\frac{[Products]}{[Reactants]}\right)$$

$$\Delta G^0 = -RT\ln K_{eq} = \Delta H^0 - T\Delta S^0$$

where ΔG is the free energy change, R is the gas constant, T is the absolute temperature, K_{eq} is the equilibrium constant, ΔG^0, ΔH^0, and ΔS^0 are the standard free energy, standard enthalpy, and standard entropy changes, respectively.

Van't Hoff equation, with a term for the heat capacity change on ligand binding

$$\ln \frac{K_1}{K_2} = \frac{\Delta H_1 - T_1 \Delta C_p}{R} \left(\frac{1}{T_1} - \frac{1}{T_2} \right) + \frac{\Delta C_p}{R} \ln \frac{T_2}{T_1}$$

where the subscripts 1 and 2 relate to different temperatures and the values for K are dissociation constants, T represents the absolute temperatures, ΔH_1 is the enthalpy of ligand binding at T_1, ΔC_p is the heat capacity of ligand binding, and R is the gas constant.

Eyring equation

$$k = \frac{k_B T}{h} e^{\left(\frac{\Delta S}{R} - \frac{\Delta H}{RT} \right)}$$

where k is the observed rate constant, k_B is the Boltzmann constant, h is Planck's constant, T is the absolute temperature, R is the gas constant, ΔH is the enthalpy of activation, and ΔS is the entropy of activation.

B.5 Other Useful Formulas

% bound

$$[\%\text{M bound}] = 100 \left(\frac{[M]_t + [L]_t + K_d - \sqrt{([M]_t + [L]_t + K_d)^2 - 4([M]_t [L]_t)}}{2[M]_t} \right)$$

where %M *bound* is the percentage of the total protein bound by ligand, $[M]_t$ is the total protein, $[L]_t$ is the total ligand, K_d is the equilibrium dissociation constant.

$$[\%\text{L bound}] = 100 \left(\frac{[M]_t + [L]_t + K_d - \sqrt{([M]_t + [L]_t + K_d)^2 - 4([M]_t [L]_t)}}{2[L]_t} \right)$$

where %L *bound* is the percentage of the total ligand bound to protein, $[M]_t$ is the total protein, $[L]_t$ is the total ligand, and K_d is the equilibrium dissociation constant.

Boltzmann thermal unfolding

$$Y = S_{min} + \frac{(S_{max} - S_{min})}{\left(1 + e^{\left(\frac{T_m - T}{slope} \right)} \right)}$$

where Y is the measured signal, S_{min} is the minimum signal, S_{max} is the maximum signal, T_m is the mid-point, T is the experimental temperature, and slope is the slope of the curve at the mid-point.

Ligand efficiency

$$LE = \frac{\Delta G}{N_h}$$

where LE is the ligand efficiency, ΔG is the Gibbs free energy of binding, and N_h is the number of heavy (non-H) atoms.

Lipophilic ligand efficiency

$$\text{LiPE} = \text{pIC}_{50} - \log P$$

where LiPE is the lipophilic ligand efficiency, pIC_{50} is the negative logarithm of the IC_{50}, and $\log P$ is the logarithm of the partition coefficient.

Appendix C

Constants, Prefixes, Conversions

C.1 Useful Physical Constants

Constant	Symbol	Value
Atomic mass unit	u	1.66054×10^{-27} kg
Avogadro's number	N_A	6.02214×10^{23} mol^{-1}
Absolute zero	$0\,K$	$-273.15\,°C$
Boltzmann constant	k	1.38066×10^{-23} JK^{-1}
Gas constant	R	8.31451 JK^{-1} mol^{-1}
Planck's constant	h	6.62608×10^{-34} Js
Standard atmospheric pressure	1 atm	1.01325×10^{5} Pa

C.2 Prefixes Used in the SI System

Prefix	Symbol	Meaning	Exponential Notation
exa	E	1,000,000,000,000,000,000	10^{18}
peta	P	1,000,000,000,000,000	10^{15}
tera	T	1,000,000,000,000	10^{12}
giga	G	1,000,000,000	10^{9}
mega	M	1,000,000	10^{6}
kilo	k	1,000	10^{3}
hector	h	100	10^{2}
deka	da	10	10^{1}
-	-	1	10^{0}

(Continued)

Laboratory Guide to Enzymology, First Edition. Geoffrey A. Holdgate, Antonia Turberville, and Alice Lanne.
© 2024 John Wiley & Sons, Inc. Published 2024 by John Wiley & Sons, Inc.

Prefix	Symbol	Meaning	Exponential Notation
deci	d	0.1	10^{-1}
centi	c	0.01	10^{-2}
milli	m	0.001	10^{-3}
micro	m	0.000001	10^{-6}
nano	n	0.000000001	10^{-9}
pico	p	0.000000000001	10^{-12}
femto	f	0.000000000000001	10^{-15}
atto	a	0.000000000000000001	10^{-18}

C.3   Conversion Factors

Length

SI unit = meter (m)

$1\,m = 100\,cm = 1000\,mm = 10^6\,mm = 10^9\,nm$

$1\,\text{Å} = 10^{-10}\,m = 10^{-8}\,cm = 10^{-1}\,nm$

Area

SI unit = square meter (m^2)

$1\,m^2 = 10^4\,cm^2$

Volume

SI unit = cubic meter (m^3)

$1\,L = 1000\,cm^3 = 10^{-3}\,m^3 = 1\,dm^3$

Mass

SI unit = kilogram (kg)

$1\,kg = 1000\,g$

$1\,\text{metric ton} = 1000\,kg$

Time

SI unit = second (s)

$1\,min = 60\,s$

$1\,hr = 3600\,s$

$1\,day = 86{,}400\,s$

$1\,year = 3.156 \times 10^7\,s$

Temperature

SI unit = Kelvin (K)

$0\,K = -273.15\,°C$

Pressure

SI unit = pascal (Pa)

$1\,Pa = 1\,N/m^2$

$1\,bar = 10^5\,Pa$

$1\,atm = 1.01325\,bar$

Energy

SI unit = joule (J)

$1\,J = 0.239\,cal$

$1\,cal = 4.184\,J$

Power

SI unit = watt (W)

$1\,W = 1\,J/s$

Radioactivity

SI unit = becquerel (Bq)

$1\,Ci = 3.7 \times 10^{10}\,Bq$

$1\,Bq = 2.7 \times 10^{-11}\,Ci$

Density

SI unit = $kg\,m^{-3}$

$1\,g\,cm^{-3} = 10^{-3}\,kg\,m^{-3}$

Viscosity

SI unit = $N\,s\,m^{-2} = kg\,m^{-1}\,s^{-1}$

C.4 Greek Alphabet

A α alpha
B β beta
Γ γ gamma
Δ δ delta
E ε epsilon
Z ζ zeta
H η eta
Θ θ theta

I ι iota
K κ kappa
Λ λ lambda
M μ mu
N ν nu
Ξ ξ xi
O o omicron
Π π pi

P ρ rho
Σ σ sigma
T τ tau
Υ υ upsilon
Φ φ phi
X χ chi
Ψ ψ psi
Ω ω omega

C.4 Greek Alphabet

A α alpha	I ι iota	P ρ rho
B β beta	K κ kappa	Σ σ sigma
Γ γ gamma	Λ λ lambda	T τ tau
Δ δ delta	M μ mu	Υ υ upsilon
E ε epsilon	N ν nu	Φ φ phi
Z ζ zeta	Ξ ξ xi	X χ chi
H η eta	O o omicron	Ψ ψ psi
Θ θ theta	Π π pi	Ω ω omega

Appendix D

Common Symbols and Abbreviations and Their Units

CHAPTER MENU

Mechanistic enzymology is, by its very nature, based on equations derived to explain the physical behavior of the enzymes under study. The symbols, abbreviations, and nomenclature used often are not obvious to the inexperienced enzymologist. Presented here is an explanation of some of the symbols used commonly in the equations used in mechanistic enzymology, along with the units that should be used for each parameter. Please note, we have used both examples where one unit is divided by another, describing this either as a division denoted by / or sometimes, for clarity, denoting the division by raising the dividing unit to the power of −1.

D.1 Common Symbols

1°	Primary
2°	Secondary
3°	Tertiary
4°	Quaternary
α	Factor by which a parameter may be increased (if $\alpha > 1$) or decreased (if $\alpha < 1$)
f	Function (indicating a mathematical relationship between quantities)
ε	Dielectric permittivity, units: F/m
$\Delta H°$	Standard enthalpy change, units: J/mol
$\Delta S°$	Standard entropy change, units: J/mol/K
$\Delta G°$	Standard Gibbs free energy change (relates thermodynamics to binding affinity, $\Delta G° = \Delta H° - TDS° = -RT\ln K_a = RT\ln K_d$), units: J/mol
ΔG^{++}	The free energy of activation, units: J/mol
[A]	Concentration of free activator, units: M
[E]	Concentration of free enzyme, units: M
[E]$_t$	Total concentration of enzyme (sum of free in solution and bound to ligands), units: M
[EA]	Concentration of enzyme-activator complex, units: M
[EI]	Concentration of enzyme-inhibitor complex, units: M

Laboratory Guide to Enzymology, First Edition. Geoffrey A. Holdgate, Antonia Turberville, and Alice Lanne.
© 2024 John Wiley & Sons, Inc. Published 2024 by John Wiley & Sons, Inc.

[ES]	Concentration of enzyme-substrate complex, units: M
[I]	Concentration of free inhibitor, units: M
[L]	Concentration of free ligand, units: M
[M]	Concentration of free macromolecule, units: M
[ML]	Concentration of macromolecule-ligand complex, units: M
[P]	Free product concentration, units: M
[S]	Free substrate concentration (not bound by enzyme), units: M
B_{max}	The maximum possible concentration of ligand-bound, units: M
C_p	Heat capacity, units: J/mol/K
D	Dielectric constant, ratio, no units
E_a	Activation energy (free energy difference between reactant and transition state), units: J/mol
h	Hill coefficient, number, no units
IC_{50}	Concentration of inhibitor giving 50% inhibition, lower number: more potent compound, units: M
k	Rate constant (note lower case k refers to a rate constant, uppercase K refers to equilibrium dissociation constant), units vary depending on reaction order, for example, a zero-order rate constant has units of Ms^{-1}, first-order rate constant has units of s^{-1}, 2nd order rate constant has units of $M^{-1}s^{-1}$
K_a	Equilibrium association constant ($= 1/K_d$), higher number: more tightly bound, units: M^{-1}
k_B	Boltzmann constant, units: J/K
k_{cat}	Catalytic constant or turnover number (number of substrate molecules converted to products per active site per unit time), units: s^{-1}
k_{cat}/K_m	Specificity constant (used to compare efficiencies of enzymes or substrates, higher number: more efficient), units: $M^{-1}s^{-1}$
K_d	Equilibrium dissociation constant (does not vary with [S] or [P]), lower number: more tightly bound, units: M
K_i	Inhibition constant, units: M
K_{i*}	Final steady-state inhibition constant (used in slow-binding kinetics), units: M
K_i'	Apparent inhibition constant (may vary with [S] or [P]), units: M
K_{ii}	Inhibition constant for intercept of Lineweaver–Burk plots (K_i' extrapolated to saturating [S]), units: M
k_{inact}	Inactivation rate constant, units vary depending on reaction order
K_{is}	Inhibition constant for slope of Lineweaver–Burk plots (K_i' extrapolated to zero [S]), units: M
K_m	Henri–Michaelis–Menten constant (concentration of substrate giving half-maximal rate), units: M
K_m^A	K_m for substrate A, units: M
k_n	Rate constant in the forward direction for step n (where n represents the number of each of the consecutive steps in the forward direction), units vary depending on reaction order
k_{-n}	Rate constant in the reverse direction for step n (where n represents the number of each of the consecutive steps in the reverse direction), units vary depending on reaction order
k_{obs}	Observed rate constant, units vary depending on reaction order

k_{off}	The dissociation rate constant, units: s^{-1}
k_{on}	The association rate constant, units: $M^{-1} s^{-1}$
K_s	Equilibrium dissociation constant for substrate, units: M
N	Native state of the protein
r	Distance, units: m
R	Gas constant (= 8.31451 J/mol/K)
T	Temperature, units: K
$t_{1/2}$	Half-life, units: s
T_m	Thermal unfolding midpoint, units: K
U	Unfolded state of the protein
v	Rate of enzyme catalyzed reaction, units: Ms^{-1} (Note: in this book, no units for rates are given for theoretically generated graphs)
V_{max}	Maximum rate of enzyme reaction, units: Ms^{-1}

D.2   Common Abbreviations

%CV	coefficient of variation
AIC	Akaike information criterion
AVR	assay variability ratio
BLI	biolayer interferometry
CD	circular dichroism
CE	capillary electrophoresis
DLS	dynamic light scattering
DSC	differential scanning calorimetry
DSF	differential scanning fluorimetry
FED	factorial experimental design
HTS	high throughput screening
ITC	isothermal titration calorimetry
IUPAC	International Union of Pure and Applied Chemistry
LoB	limit of blank
LoD	limit of detection
LoQ	limit of quantification
MAD	median absolute difference
MALS	multi-angle light scattering
MS	mass spectrometry
MSR	minimum significant ratio
NMR	nuclear magnetic resonance
OVAT	one variable at a time
PAINS	pan-assay interference compounds
PD	pharmacodynamics
PK	pharmacokinetics
QC	quality control
RP-HPLC	reversed-phase-high performance liquid chromatography
RWG	resonance waveguide grating
S : B	signal: background

S : N	signal: noise
SD	standard deviation
SPR	surface plasma resonance
SW	signal window
TIC	total ion chromatogram
TS	transition state
UV	ultraviolet

Appendix E

Glossary

Activation energy	The minimum energy that reactants must have in order to form products.
Activator	A species that binds to an enzyme and increases its activity. Opposite to an inhibitor.
Active site	Amino acids and side chains which are in direct physical contact with the substrate, and those side chains that may not be in direct contact, but which perform a function in the catalytic cycle.
Activity	Ability of a quantity of enzyme to catalyze a specific reaction. Often used to quantify the amount of enzyme present – 1 unit of enzyme is the amount which catalyzes the formation of 1 mmol of product per minute under specified conditions. This unit is, however, not particularly useful in mechanistic studies, and it is preferred that k_{cat} values, which have more physical meaning, are used to quantify the turnover of substrates to products.
Allostery	The ability of a binding event at one site to influence binding events at other sites. Also known as cooperativity (+ve where binding leads to an increase in affinity at subsequent sites, −ve where binding leads to a decrease in affinity at subsequent sites).
Apoenzyme	Protein part of an active species (enzyme and cofactor) for enzymes requiring cofactors for activity.
Assay	Method for determining the rate of product formation (or substrate used) during the enzyme-catalyzed reaction.
Auxiliary enzyme	Enzyme added to an assay to allow the conversion of the product of the reaction under study to another species that can be measured.
Binary complex	Complex between two interacting partners (macromolecule and ligand).
Binding	Combination of the enzyme and ligand to yield an enzyme-ligand complex.
Binding site	Amino acids and side chains involved in the interaction with ligand (compared with active site).
Burst	Rapid accumulation of an enzyme-bound intermediate, which leads to an initial rapid release of product, followed by a progressive increase as the intermediate is turned over.

Laboratory Guide to Enzymology, First Edition. Geoffrey A. Holdgate, Antonia Turberville, and Alice Lanne.
© 2024 John Wiley & Sons, Inc. Published 2024 by John Wiley & Sons, Inc.

Catalysis	Ability of an enzyme to increase the rate of a chemical reaction, without undergoing a change in itself.
Catalytic constant (k_{cat})	Limiting value of all the first-order rate constants on the forward reaction pathway. It represents the maximum number of substrate molecules turned over to product per active site per unit time. Often termed the turnover number. Units of k_{cat} should be s^{-1}.
Coenzyme	The non-protein species required by some enzymes for catalysis.
Cofactor	See Coenzyme.
Competition	Effect observed when two ligands that are mutually exclusive for binding are included in the same assay.
Competitive inhibition	Type of inhibition where inhibitor binds only before the varied substrate (see also Competition).
Cooperativity	See Allostery.
Coupled assay	Assay makes use of an auxiliary enzyme to enable the reaction of the primary enzyme to be followed by producing a measurable product.
Denaturation	Alteration of the conformation of an enzyme by physical or chemical means, often irreversibly, with a concomitant loss of catalytic activity.
Discontinuous assay	Enzyme assay in which samples are removed at one or more time points and analyzed to determine the extent of reaction.
Double reciprocal plot	Linearized plot in which the reciprocal of the bound ligand concentration (1/[B]) is plotted versus the reciprocal of the free concentration (1/[L]) for a binding reaction in order to display the values of K_d and stoichiometry (n). The slope is K_d/n and intercept is $1/n$. It should be used for display purposes not calculating constants. See also Lineweaver-Burk plot (1/v versus 1/[S]).
Enzyme	Biological catalysts, which are composed of a 3D arrangement of peptide-linked amino acids.
Enzyme assay	See Assay.
Equilibrium	State of balance in a chemical reaction, where the rate of the forward reaction equals the rate of the reverse reaction, and no net change in concentration of substrates or products occurs (see Steady state).
First-order kinetics	Reaction rate kinetics, which can be described by a constant multiplied by a concentration.
Gibbs free energy	A measure of how likely a reaction is to occur (−ve value suggests the reaction is likely to occur spontaneously if the reactants are present at a standard concentration, usually 1 M). It does not indicate how fast a reaction will occur.
Hill kinetics	Kinetics displaying cooperativity, described by the Hill equation. See also Allostery.
IC$_{50}$	The inhibitor concentration giving a 50% decrease in rate.
Inactivation	Loss of activity of an enzyme, causing a decrease in rate. See also Denaturation.
Inhibition	Decrease in the catalytic rate of an enzyme-catalyzed reaction by a bound ligand (inhibitor).

Initial rate	Rate of reaction measured at the start of an enzyme reaction (in the steady state) during which the enzyme-bound intermediate concentrations are not changing, and the accumulation of product is linear with time.
Irreversible inhibition	Inhibition that cannot be reversed on the timescale of the assay. Truly irreversible inhibitors never dissociate from the enzyme and are characterized by an inactivation rate constant, not a dissociation constant. Irreversible inhibitors do not necessarily have to be covalently bound.
Isoenzyme or isozyme	A different molecular form of the same enzyme. Isoenzymes have very similar amino acid sequences.
K_m	See Michaelis constant.
Ligand	Chemical species that binds to an enzyme.
Lineweaver-Burk plot	Double reciprocal plot ($1/v$ versus $1/[S]$) useful for displaying enzyme kinetic data. The slope is K_m/V_{max} and intercept is $1/V_{max}$. Should be used only for display purposes, not for calculation of kinetic constants. See also Double reciprocal plot.
Maximum rate (velocity)	Rate observed when the enzyme is saturated with substrate.
Mechanism	Process by which a reaction takes place. Fully determined when all the intermediates, complexes, and conformational states of an enzyme are characterized, with the rate constants associated with the conversion between them quantified. Term often used in inhibition studies to distinguish between different modes of inhibition.
Michaelis constant (K_m)	An apparent dissociation constant, the value of which can be greater or smaller than the true substrate dissociation constant. K_m is the substrate concentration that gives half the maximum rate.
Michaelis-Menten equation	Equation describing the kinetics of some simple single substrate enzyme catalyzed reactions.
Microscopic reversibility	States that the reaction pathway for a reaction at equilibrium is the exact opposite of the pathway for the reverse reaction. Once a possible reaction mechanism is found for one direction, a possible reaction mechanism is found for the reverse direction.
Mixed inhibition	Type of inhibition in which inhibitor binds both before and after the varied substrate.
Multi-substrate reaction	Reaction catalyzed in which more than a single substrate binds to the enzyme, either together or separately, and is converted to products.
Noncompetitive inhibition	Type of inhibition in which inhibitor binds with equal affinity both before and after the varied substrate.
Nonlinear regression	Method of data-fitting involving minimization of the square of the distance of the data points to the fitted curve, applied to nonlinear functions.
Optimum pH	pH value at which the rate of an enzyme-catalyzed reaction is at a maximum under specified conditions.

Optimum temperature	Temperature at which the rate of an enzyme-catalyzed reaction is at a maximum under specified conditions.
Order of reaction	A description of the kinetics of a reaction detailing the number of concentration terms multiplied together in the expression for the rate.
Partial inhibition	Type of inhibition in which binding of the inhibitor does not completely block product formation.
Ping–pong mechanism	A mechanism in which a product is released between the addition of the first substrate and the addition of the second substrate. These mechanisms involve reversible covalent modification of the enzyme.
Pre-steady state	Initial period, following mixing of the substrate, during which the concentration of enzyme-bound intermediates accumulate to their steady-state levels. The pre-steady state usually lasts only a fraction of a second.
Product inhibition	Inhibition observed when one of the products of the reaction binds to the enzyme and decreases the rate of reaction.
Progress curve	Plot of the concentration of product accumulated versus the time of reaction, also known as the time-course.
Random mechanism	Reaction mechanism in which either of the two substrates may bind to the enzyme first, followed by the other to form a ternary complex.
Rate	Change in the product concentration per unit of time, also known as velocity.
Rate constant	Constant of proportionality used along with concentration term(s) to define the rate of a reaction.
Reversible inhibition	Inhibition that can be reversed on the timescale of the assay, by competition or dilution. Reversible inhibitors are not precluded from forming covalent bonds with the enzyme.
Saturation	Condition occurring when all of the enzyme active sites are filled with substrate, leading to maximal velocity, or all ligand sites are filled leading to maximal binding.
Selwyn's test	Test to check for instability of enzyme during a reaction.
Sigmoidal kinetics	See Allostery.
Slow-binding inhibition	Inhibition which occurs slowly on the timescale of the assay, as the enzyme-inhibitor complex concentration increases to its steady state level.
Specificity	Discrimination by the enzyme for two different substrates.
Specificity constant	Numerical expression of specificity, V_{max}/K_m or more correctly k_{cat}/K_m.
Steady-state	Period during which the concentration of enzyme-bound intermediates is not changing. It is during the steady-state period that enzyme-catalyzed reaction rates are usually measured.
Substrate inhibition	Inhibition observed when higher concentrations of substrate bind to the enzyme and actually cause a decrease in reaction rate.

Ternary complex	Complex formed when two separate ligands (often substrates) bind to the enzyme simultaneously, to form a complex consisting of three species.
Tight-binding inhibition	Type of inhibition occurring when the concentration of inhibitor required to cause inhibition is similar to the enzyme concentration, leading to depletion of the free inhibitor concentration. This leads to breakdown of the usual assumptions leading to simple inhibition kinetics and requires a more complex rate equation.
Time-course	See Progress curve.
Transition state	Unstable, high-energy state occurs on a reaction pathway in which bonds are being made and broken. Enzymes bind the transition state more tightly than the ground state substrates, therefore potent inhibitors can theoretically be designed from knowledge of transition state structure.
Turnover number	See Catalytic constant.
Uncompetitive inhibition	Type of inhibition in which the inhibitor binds only after the varied substrate.
Velocity	See Rate.
V_{max}	See Maximum velocity.

Appendix F

Key Derivations

CHAPTER MENU

F.1 Langmuir Isotherm, Assuming Rapid Equilibrium

$$M + L \rightleftharpoons ML$$

$$K_d = \frac{[M][L]}{[ML]}$$

$$[M]_t = [M] + [ML]$$

$$[L] \approx [L]_t \text{ if } [L] \gg [ML]$$

$$[M]_t = [M]\left(1 + \frac{[L]}{K_d}\right)$$

$$[M] = \frac{[M]_t}{\left(1 + \frac{[L]}{K_d}\right)}$$

Laboratory Guide to Enzymology, First Edition. Geoffrey A. Holdgate, Antonia Turberville, and Alice Lanne.
© 2024 John Wiley & Sons, Inc. Published 2024 by John Wiley & Sons, Inc.

$$[ML] = [M]_t - [M]$$

$$[ML] = \left\{ [M]_t \frac{\left(1 + \frac{[L]}{K_d}\right)}{\left(1 + \frac{[L]}{K_d}\right)} \right\} - \left\{ \frac{[M]_t}{\left(1 + \frac{[L]}{K_d}\right)} \right\}$$

$$[ML] = [M]_t \frac{\left(\frac{[L]}{K_d}\right)}{\left(1 + \frac{[L]}{K_d}\right)}$$

$$[ML] = \frac{[M]_t[L]}{K_d + [L]}$$

Which is Equation 2.16.

F.2 Michaelis-Menten, Assuming Rapid Equilibrium

$$E + S \underset{}{\overset{K_m}{\rightleftharpoons}} ES \xrightarrow{k_{cat}} E + P$$

$$[E]_t = [E] + [ES]$$

$$K_m = \frac{[E][S]}{[ES]}$$

$$v = k_{cat}[ES]$$

$$V_{max} = k_{cat}[E]_t$$

$$\frac{v}{V_{max}} = \frac{[ES]}{[E]_t} = \frac{[ES]}{([E] + [ES])}$$

$$\frac{v}{V_{max}} = \frac{\left(\frac{[E][S]}{K_m}\right)}{\left([E] + \frac{[E][S]}{K_m}\right)}$$

$$\frac{v}{V_{max}} = \frac{\left(\frac{[S]}{K_m}\right)}{\left(1 + \frac{[S]}{K_m}\right)}$$

$$\frac{v}{V_{max}} = \frac{[S]}{(K_m + [S])}$$

$$v = \frac{V_{max}[S]}{(K_m + [S])} = \frac{k_{cat}[E]_t[S]}{K_m + [S]}$$

Which is Equation 5.4.

F.3   Michaelis-Menten, Assuming Steady-state

$$E + S \underset{k_{-1}}{\overset{k_1}{\rightleftharpoons}} ES \xrightarrow{k_{cat}} E + P$$

$$\frac{d[E]}{dt} = 0 = (k_{cat} + k_{-1})[ES] - k_1[E][S]$$

$$[ES] = \frac{k_1[E][S]}{(k_{-1} + k_{cat})}$$

$$v = k_{cat}[ES] = \frac{k_{cat}k_1[E][S]}{(k_{-1} + k_{cat})}$$

$$[E]_t = [E] + [ES] = [E] + \frac{k_1[E][S]}{(k_{-1} + k_{cat})}$$

$$[E] = \frac{[E]_t}{\left(1 + \frac{k_1[S]}{(k_{-1}+k_{cat})}\right)}$$

$$v = k_{cat}[ES] = \frac{k_{cat}k_1[E][S]}{(k_{-1} + k_{cat})}$$

$$v = \frac{k_{cat}k_1[S]}{(k_{-1} + k_{cat})} \cdot \frac{[E]_t}{\left(1 + \frac{k_1[S]}{(k_{-1}+k_{cat})}\right)}$$

$$v = \frac{k_{cat}k_1[E]_t[S]}{((k_{-1} + k_{cat}) + k_1[S])}$$

$$v = \frac{k_{cat}[E]_t[S]}{\left(\frac{(k_{-1}+k_{cat})}{k_1} + [S]\right)}$$

$$v = \frac{V_{max}[S]}{(K_m + [S])}$$

Which is Equation 5.4, where $V_{max} = k_{cat}[E]_t$, and

$$K_m = \frac{(k_{-1} + k_{cat})}{k_1}$$

F.4   Rate is Directly Proportional to Free Enzyme

From Equation 5.7, K_m is given by:

$$K_m = \frac{[E][S]}{\sum[ES]}$$

Substituting into Equation 5.4

$$v = \frac{k_{cat}[E]_t[S]}{K_m + [S]}$$

$$[E]_t = \sum [ES] + [E]$$

$$[E]_t = \frac{[E][S]}{K_m} + [E]$$

$$[E]_t = [E] \left(\frac{K_m + [S]}{K_m} \right)$$

$$v = \frac{k_{cat}[S][E] \left(\frac{K_m + [S]}{K_m} \right)}{K_m + [S]}$$

$$v = \frac{k_{cat}[S][E]}{K_m}$$

Which is Equation 5.6.

F.5 Two Substrate Reactions

$$v = k_{cat}[EAB]$$

$$\frac{v}{[E]_t} = \frac{k_{cat}[EAB]}{[E] + [EA] + [EB] + [EAB]}$$

$$v = \frac{V_{max}[EAB]}{[E] + [EA] + [EB] + [EAB]}$$

$$[E] = \frac{K_a[EAB]K_m{}^B}{[A][B]}$$

$$[EA] = \frac{K_m{}^B[EAB]}{[B]}$$

$$[EB] = \frac{K_m{}^A[EAB]}{[A]}$$

$$v = \frac{V_{max}[EAB]}{\left(\frac{K_a[EAB]K_m{}^B}{[A][B]} + \frac{K_m{}^B[EAB]}{[B]} + \frac{K_m{}^A[EAB]}{[A]} + [EAB] \right)}$$

Dividing by [EAB] gives

$$v = \frac{V_{max}}{\left(1 + \frac{K_m^{\ A}}{[A]} + \frac{K_m^{\ B}}{[B]} + \frac{K_a K_m^{\ B}}{[A][B]}\right)}$$

Which is Equation 5.11.

F.6 Dose-response Equation to Calculate K_i'

For classical kinetics (i.e. not tight binding), we can assume that $[I]_t \gg [I]_{bound}$, so that $[I]_t = [I]_f$, and that $IC_{50} = K_i'$.

$$E' + I \xrightleftharpoons{K_i'} EI$$

Rate in the absence of I is $k_{cat}[E]_t$:

$$v_0 = k_{cat}[E]_t$$

Rate in the presence of I is $k_{cat}[E']$, where $[E']$ is Enzyme not bound by I:

$$v_i = k_{cat}[E']$$

$$K_i' = \frac{[E'][I]}{[E'I]}$$

$$[E]_t = [E'] + [E'I]$$

$$[E]_t = [E'] + \left(\frac{[E'][I]}{K_i'}\right)$$

$$[E]_t = [E']\left(1 + \frac{[I]}{K_i'}\right)$$

$$\frac{v_i}{v_0} = \frac{[E']}{[E]_t} = \frac{1}{\left(1 + \frac{[I]}{K_i'}\right)}$$

$$v_i = \frac{v_0}{\left(1 + \frac{[I]}{K_i'}\right)}$$

Which is Equation 6.2.

F.7 Derivation of Substrate Inhibition, Assuming Rapid Equilibrium

$$E + S \xrightleftharpoons{K_m} ES \xrightarrow{k_{cat}} E + P$$

$$+$$
$$S$$

$$ES_2$$

$$[E]_t = [E] = [ES] + [ES_2]$$

$$K_m = \frac{[E][S]}{[ES]}$$

$$K_{si} = \frac{[ES][S]}{[ES_2]}$$

$$\frac{v}{V_{max}} = \frac{k_{cat}[ES]}{k_{cat}[E]_t} = \frac{[ES]}{[E] + [ES] + [ES_2]}$$

$$v = \frac{V_{max}\frac{[E][S]}{K_m}}{[E] + \frac{[E][S]}{K_m} + \frac{[E][S][S]}{K_m K_{si}}}$$

$$v = \frac{V_{max}[S]}{K_m + [S] + \frac{[S]^2}{K_{si}}}$$

Which is Equation 5.8.

F.8 Competing Ligands, Assuming Rapid Equilibrium

Consider the scheme:

By conservation of mass:

$$[M]_t = [M] + [ML] + [MA]$$

$$K_{dL} = \frac{[M][L]}{[ML]}$$

$$K_{dA} = \frac{[M][A]}{[MA]}$$

$$\frac{[B]}{B_{max}} = \frac{[ML]}{B_{max}} = \frac{[ML]}{[M] + [MA] + [ML]}$$

$$\frac{[B]}{B_{max}} = \frac{\frac{[M][L]}{K_{dL}}}{[M] + \frac{[M][A]}{K_{dA}} + \frac{[M][L]}{K_{dL}}}$$

$$[B] = \frac{B_{max} [L]}{K_{dL} \left(1 + \frac{[A]}{K_{dA}}\right) + [L]}$$

$$[B] = \frac{[M]_t [L]}{K_{dL} \left(1 + \frac{[A]}{K_{dA}}\right) + [L]}$$

Which is Equation 2.27.

F.9 Tight Binding

The phenomenon of tight-binding kinetics occurs when there is depletion of the free ligand (often inhibitor) concentration, caused by binding to the enzyme, so that the usual assumption that the total concentration equals free concentration is not valid (i.e. $[I]_t \neq [I]_f$).

The dose-response equation is derived as follows:

Rate in the absence of I is $k_{cat}[E]_t$:

$$v_0 = k_{cat}[E]_t$$

Rate in the presence of I is $k_{cat}[E']$, where $[E']$ is Enzyme not bound by I; $v_i = k_{cat}[E']$

$$v_i = k_{cat}[E']$$

$$K_i' = \frac{[E'][I]}{[E'I]}$$

$$[E]_t = [E'] + [E'I]$$

$$[I]_t = [I] + [E'I]$$

$$K_i' = [E']\frac{([I]_t - [E'I])}{([E]_t - [E'I])}$$

$$K_i' = [E']\frac{([I]_t - [E]_t + [E'])}{([E]_t - [E'])}$$

$$K_i'[E]_t - K_i'[E'] = [E'][I]_t - [E'][E]_t + [E']^2$$

$$[E']^2 + [E']\left([I]_t - [E]_t + K_i'\right) - K_i'[E]_t = 0$$

$$[E'] = \frac{-\left([I]_t - [E]_t + K_i'\right) + \sqrt{\left([I]_t - [E]_t + K_i'\right)^2 - 4K_i[E]_t}}{2}$$

$$\frac{v_i}{v_0} = \frac{[E']}{[E]_t}$$

$$v_i = v_0 \cdot \frac{-\left([I]_t - [E]_t + K_i'\right) + \sqrt{\left([I]_t - [E]_t + K_i'\right)^2 - 4K_i[E]_t}}{2[E]_t}$$

This can be expressed as:

$$v_i = 0.5 \cdot v_0 \cdot \left\{ -\left(\frac{K_i'}{[E]_t} + \frac{[I]_t}{[E]_t} - 1\right) + \sqrt{\left(\frac{K_i'}{[E]_t} + \frac{[I]_t}{[E]_t} - 1\right)^2 + \frac{4K_i'}{[E]_t}} \right\}$$

Which is Equation 6.7.

F.10 Single Exponential, with First-order Rate Equation

Consider the scheme

$$A \xrightarrow{\quad k \quad} B$$

$$\frac{d[A]}{dt} = -k[A]$$

and

$$\frac{d[B]}{dt} = k[A]$$

Separating the variables

$$\int \frac{1}{[A]} d[A] = \int -k \, dt$$

$$\ln[A] + c = -kt$$

At $t = 0$, $[A] = [A]_0$

$$\ln[A]_0 + c = 0$$

so

$$c = -\ln[A]_0$$

$$\ln[A] - \ln[A]_0 = -kt$$

$$\ln \frac{[A]}{[A]_0} = -kt$$

$$\frac{[A]}{[A]_0} = e^{-kt}$$

$$[A] = [A]_0 e^{-kt}$$

which is Equation 2.4.

F.11   Protein Double Ionization

Consider the scheme:

$$H_2A \xrightleftharpoons{K_1} HA^- + H^+ \xrightleftharpoons{K_2} A^{2-} + 2H^+$$

$$K_1 = \frac{[HA][H]}{[H_2A]}$$

$$K_2 = \frac{[A][H]}{[HA]}$$

$$[A]_0 = [H_2A] + [HA] + [A]$$

Expressing $[A]_0$ in terms of $[HA]$:

$$[A]_0 = \frac{[HA][H]}{K_1} + [HA] + \frac{K_2[HA]}{[H]}$$

$$[A]_0 = [HA]\left(1 + \frac{[H]}{K_1} + \frac{K_2}{[H]}\right)$$

Expressing $[A]_0$ in terms of $[A]$:

$$[A]_0 = [A] + \frac{[A][H]}{K_2} + \frac{[A][H][H]}{K_1K_2}$$

$$[A]_0 = [A]\left(1 + \frac{[H]}{K_2} + \frac{[H]^2}{K_1K_2}\right)$$

Expressing $[A]_0$ in terms of $[H_2A]$:

$$[A]_0 = [H_2A] + \frac{K_1[H_2A]}{[H]} + \frac{K_1K_2[H_2A]}{[H][H]}$$

$$[A]_0 = [H_2A]\left(1 + \frac{K_1}{[H]} + \frac{K_1K_2}{[H]^2}\right)$$

Now

$$L_H[A]_0 = L_{H_2A}[H_2A] + L_{HA}[HA] + L_A[A]$$

where

$$[HA] = \frac{[A]_0}{\left(1 + \frac{[H]}{K_1} + \frac{K_2}{[H]}\right)}$$

$$[A] = \frac{[A]_0}{\left(1 + \frac{[H]}{K_2} + \frac{[H]^2}{K_1K_2}\right)}$$

$$[H_2A] = \frac{[A]_0}{\left(1 + \frac{K_1}{[H]} + \frac{K_1K_2}{[H]^2}\right)}$$

$$L_H[A]_0 = \frac{L_{H_2A}[A]_0}{\left(1 + \frac{K_1}{[H]} + \frac{K_1K_2}{[H]^2}\right)} + \frac{L_{HA}[A]_0}{\left(1 + \frac{[H]}{K_1} + \frac{K_2}{[H]}\right)} + \frac{L_A[A]_0}{\left(1 + \frac{[H]}{K_2} + \frac{[H]^2}{K_1K_2}\right)}$$

Dividing by $[A]_0$, and multiplying through to collect terms in the denominator:

$$L_H = \frac{L_{H_2A}[H]^2}{([H]^2 + K_1[H] + K_1K_2)} + \frac{L_{HA}K_1[H]}{(K_1[H] + [H]^2 + K_1K_2)} + \frac{L_AK_1K_2}{(K_1K_2 + K_1[H] + [H]^2)}$$

Simplifying:

$$L_H = \frac{L_{H_2A}[H]^2 + L_{HA}K_1[H] + L_AK_1K_2}{K_1K_2 + K_1[H] + [H]^2}$$

Which is Equation 4.21.

When $L_{H_2A} = L_A = 0$, then

$$L_H = \frac{L_{HA}}{1 + \frac{[H]}{K_1} + \frac{K_2}{[H]}}$$

Which is Equation 4.23.

F.12   Protein Single Ionization

Consider the scheme:

$$H_2A \underset{}{\overset{K_a}{\rightleftharpoons}} HA^- + H^+$$

$$[A]_0 = [HA] + [A]$$

$$[A]_0 = [HA] + \frac{[HA]K_a}{[H]}$$

$$[A]_0 = [HA]\left(1 + \frac{K_a}{[H]}\right)$$

$$[HA] = \frac{[A]_0}{1 + \frac{K_a}{[H]}}$$

$$[A]_0 = [HA] + [A]$$

$$[A]_0 = \frac{[A][H]}{K_1} + [A]$$

$$[A]_0 = [A]\left(1 + \frac{[H]}{K_a}\right)$$

$$[A] = \frac{[A]_0}{1 + \frac{[H]}{K_a}}$$

$$L_H[A]_0 = L_{HA}[HA] + L_A[A]$$

$$L_H[A]_0 = \frac{L_{HA}[A]_0}{1 + \frac{K_a}{[H]}} + \frac{L_A[A]_0}{1 + \frac{[H]}{K_a}}$$

$$L_{\mathrm{H}} = \frac{L_{\mathrm{HA}}[\mathrm{H}]}{[\mathrm{H}] + K_{\mathrm{a}}} + \frac{L_{\mathrm{A}}K_{\mathrm{a}}}{K_{\mathrm{a}} + [\mathrm{H}]}$$

$$L_{\mathrm{H}} = \frac{L_{\mathrm{HA}}[\mathrm{H}] + L_{\mathrm{A}}K_{\mathrm{a}}}{K_{\mathrm{a}} + [\mathrm{H}]}$$

Which is Equation 4.24.

F.13 Derivation of the Integrated Rate Equation for Slow-binding Inhibition Described by a Two-step Mechanism

Consider the scheme:

The kinetic scheme for inhibition following a two-step mechanism is shown in the text. The enzyme initially partitions between E, EI and ES. At steady-state, EI* also exists. During the shift from initial state to steady-state.

$$\frac{\mathrm{d}}{\mathrm{d}t}([\mathrm{E}] + [\mathrm{ES}] + [\mathrm{EI}]) = -k_5[\mathrm{EI}] + k_6[\mathrm{EI}^*]$$

$$K_{\mathrm{m}} = \frac{[\mathrm{E}][\mathrm{S}]}{[\mathrm{ES}]}$$

$$[\mathrm{E}] = \frac{[\mathrm{ES}]K_{\mathrm{m}}}{[\mathrm{S}]}$$

$$K_{\mathrm{i}} = \frac{[\mathrm{E}][\mathrm{I}]}{[\mathrm{EI}]}$$

$$[\mathrm{EI}] = \frac{[\mathrm{E}][\mathrm{I}]}{K_{\mathrm{i}}}$$

$$[\mathrm{EI}] = [\mathrm{ES}]\left(\frac{K_{\mathrm{m}}}{[\mathrm{S}]}\right)\left(\frac{[\mathrm{I}]}{K_{\mathrm{i}}}\right)$$

Substituting for [E] and [EI] gives:

$$\frac{\mathrm{d}}{\mathrm{d}t}[\mathrm{ES}]\left(1 + \left(\frac{K_{\mathrm{m}}}{[\mathrm{S}]}\right) + \left(\frac{K_{\mathrm{m}}}{[\mathrm{S}]}\right)\left(\frac{[\mathrm{I}]}{K_{\mathrm{i}}}\right)\right) = -k_5[\mathrm{EI}] + k_6[\mathrm{EI}^*]$$

$$[\mathrm{EI}^*] = \frac{\mathrm{d}}{\mathrm{d}t}[\mathrm{ES}]\frac{1}{k_6}\left(1 + \left(\frac{K_{\mathrm{m}}}{[\mathrm{S}]}\right) + \left(\frac{K_{\mathrm{m}}}{[\mathrm{S}]}\right)\left(\frac{[\mathrm{I}]}{K_{\mathrm{i}}}\right)\right) + \frac{k_5}{k_6}[\mathrm{ES}]\left(\frac{K_{\mathrm{m}}}{[\mathrm{S}]}\right)\left(\frac{[\mathrm{I}]}{K_{\mathrm{i}}}\right)$$

By conservation of mass, the total concentration of enzyme is given by

$$[E]_t = [E] + [ES] + [EI] + [EI^*]$$

Substituting for $[E]$, $[EI]$, and $[EI^*]$ gives:

$$[E]_t = [ES]\left(1 + \left(\frac{K_m}{[S]}\right) + \left(\frac{K_m}{[S]}\right)\left(\frac{[I]}{K_i}\right)\right) + \frac{d}{dt}[ES]\frac{1}{k_6}\left(1 + \left(\frac{K_m}{[S]}\right) + \left(\frac{K_m}{[S]}\right)\left(\frac{[I]}{K_i}\right)\right)$$
$$+ \frac{k_5}{k_6}[ES]\left(\frac{K_m}{[S]}\right)\left(\frac{[I]}{K_i}\right)$$

$$\frac{d}{dt}[ES]\frac{1}{k_6}\left(1 + \left(\frac{K_m}{[S]}\right) + \left(\frac{K_m}{[S]}\right)\left(\frac{[I]}{K_i}\right)\right) = [E]_t - [ES]\left(1 + \left(\frac{K_m}{[S]}\right) + \left(\frac{K_m}{[S]}\right)\left(\frac{[I]}{K_i}\right)\right.$$
$$\left. + \frac{k_5}{k_6}\left(\frac{K_m}{[S]}\right)\left(\frac{[I]}{K_i}\right)\right)$$

$$\frac{d}{dt}[ES] = \frac{k_6[E]_t}{\left(1 + \left(\frac{K_m}{[S]}\right) + \left(\frac{K_m}{[S]}\right)\left(\frac{[I]}{K_i}\right)\right)} - \frac{k_6[ES]\left(1 + \left(\frac{K_m}{[S]}\right) + \left(\frac{K_m}{[S]}\right)\left(\frac{[I]}{K_i}\right) + \frac{k_5}{k_6}\left(\frac{K_m}{[S]}\right)\left(\frac{[I]}{K_i}\right)\right)}{\left(1 + \left(\frac{K_m}{[S]}\right) + \left(\frac{K_m}{[S]}\right)\left(\frac{[I]}{K_i}\right)\right)}$$

Simplifying:

$$\frac{d}{dt}[ES] = b - k[ES]$$

where

$$b = \frac{k_6[E]_t}{\left(1 + \left(\frac{K_m}{[S]}\right) + \left(\frac{K_m}{[S]}\right)\left(\frac{[I]}{K_i}\right)\right)}$$

and

$$k = \frac{k_6\left(1 + \left(\frac{K_m}{[S]}\right) + \left(\frac{K_m}{[S]}\right)\left(\frac{[I]}{K_i}\right) + \frac{k_5}{k_6}\left(\frac{K_m}{[S]}\right)\left(\frac{[I]}{K_i}\right)\right)}{\left(1 + \left(\frac{K_m}{[S]}\right) + \left(\frac{K_m}{[S]}\right)\left(\frac{[I]}{K_i}\right)\right)}$$

so

$$k = k_6 + \left[\frac{k_5\left(\frac{[I]}{K_i}\right)}{\left(1 + \left(\frac{[S]}{K_m}\right) + \left(\frac{[I]}{K_i}\right)\right)}\right]$$

From above

$$dt = \frac{d[ES]}{(b - k[ES])}$$

Integration of this standard form gives

$$t + c = -\frac{1}{a}\ln(b - k[ES])$$

When $t = 0$, I behaves as a competitive inhibitor with respect to S

$$[ES] = \frac{[E]_t[S]}{K_m\left(1 + \left(\frac{[I]}{K_i}\right)\right) + [S]}$$

Substituting

$$c = -\frac{1}{k} \ln \left[b - \frac{k[E]_t[S]}{K_m \left(1 + \left(\frac{[I]}{K_i}\right)\right) + [S]} \right]$$

So

$$t - \frac{1}{k} \ln \left[b - \frac{k[E]_t[S]}{K_m \left(1 + \left(\frac{[I]}{K_i}\right)\right) + [S]} \right] = -\frac{1}{k} \ln(b - k[ES])$$

$$-kt = \ln \left[\frac{(b - k[ES])}{b - \frac{k[E]_t[S]}{K_m\left(1 + \left(\frac{[I]}{K_i}\right)\right)}} \right]$$

$$-kt = \ln \left[\frac{(b/k - [ES])}{b/k - \frac{[E]_t[S]}{K_m\left(1 + \left(\frac{[I]}{K_i}\right)\right) + [S]}} \right]$$

At steady-state, $d[ES]/dt = 0$, so $[ES]_s = b/k$, and

$$-kt = \ln \left[\frac{([ES]_s - [ES])}{[ES]_s - \frac{[E]_t[S]}{K_m\left(1 + \left(\frac{[I]}{K_i}\right)\right) + [S]}} \right]$$

The rate of reaction, $v = k_{cat}[ES]$, so

$$-kt = \ln \left[\frac{v_s - v}{v_s - v_0} \right]$$

where the subscripts s and 0 relate to steady and initial states, respectively.
Thus,

$$e^{-kt} = \frac{(v_s - v)}{(v_s - v_0)}$$

$$(v_s - v_0)e^{-kt} = (v_s - v)$$

So

$$v = v_s + (v_0 - v_s)e^{-kt}$$

Rate is $d[P]/dt$, so that

$$\int d[P] = v_s \int dt + (v_0 - v_s) \int e^{-kt}dt$$

So

$$[P] = v_s t - \left(\frac{v_0 - v_s}{a}\right)e^{-kt} + c$$

When $t = 0$, $[P] = 0$, so that

$$c = \frac{(v_0 - v_s)}{k}$$

So

$$[P] = v_s t + \frac{(v_0 - v_s)(1 - e^{-kt})}{k}$$

Which is Equation 6.9 in the text.

Also at steady-state,

$$\frac{v}{[E]_t} = \frac{k_{cat}[ES]}{[E] + [ES] + [EI] + [EI^*]}$$

$$\frac{v}{[E]_t} = \frac{k_{cat}[S]/K_m}{1 + \left(\frac{[S]}{K_m}\right) + \left(\frac{[I]}{K_i^*}\right)}$$

And

$$v = \frac{V_{max}[S]}{K_m \left(1 + \left(\frac{[I]}{K_i^*}\right)\right) + [S]}$$

This equation is the same as Equation 6.13 in the text and describes the variation in the steady-state rate with [I].

F.14 Essential Activation

Consider the scheme:

$$v = k_{cat}[EAS]$$

$$\frac{v}{[E]_t} = \frac{k_{cat}[EAS]}{[E] + [EA] + [ES] + [EAS]}$$

$$\frac{v}{[E]_t} = \frac{k_{cat}\frac{[E][A][S]}{K_x \alpha K_s}}{[E] + \frac{[E][A]}{K_x} + \frac{[E][S]}{K_s} + \frac{[E][A][S]}{K_x \alpha K_s}}$$

Divide top and bottom by $\frac{[E][A]}{K_x \alpha K_s}$

$$\frac{v}{[E]_t} = \frac{k_{cat}[S]}{\frac{K_x \alpha K_s}{[A]} + \alpha K_s + \frac{[S]\alpha K_x}{[A]} + [S]}$$

$$v = \frac{V_{max}[S]}{\alpha K_s \left(1 + \frac{K_x}{[A]}\right) + [S]\left(1 + \frac{\alpha K_x}{[A]}\right)}$$

Which is Equation 7.1.

F.15   Non-essential Activation

Consider the scheme:

$$\text{E} \;+\; \text{S} \underset{}{\overset{K_\text{m}}{\rightleftharpoons}} \text{ES} \xrightarrow{k_\text{cat}} \text{E} + \text{P}$$

$$+ \qquad\qquad\qquad +$$

$$\text{A} \qquad\qquad\qquad \text{A}$$

$$\Big\updownarrow K_\text{x} \qquad\qquad \Big\updownarrow \alpha K_\text{x}$$

$$\text{EA} \;+\; \text{S} \underset{}{\overset{\alpha K_\text{m}}{\rightleftharpoons}} \text{EAS} \xrightarrow{\beta k_\text{cat}} \text{EA} + \text{P}$$

$$v = k_\text{cat}[\text{ES}] + \beta k_\text{cat}[\text{EAS}]$$

$$\frac{v}{[\text{E}]_\text{t}} = \frac{k_\text{cat}[\text{ES}] + \beta k_\text{cat}[\text{EAS}]}{[\text{E}] + [\text{EA}] + [\text{ES}] + [\text{EAS}]}$$

$$\frac{v}{[\text{E}]_\text{t}} = \frac{k_\text{cat}\frac{[\text{E}][\text{S}]}{K_\text{m}} + \beta k_\text{cat}\frac{[\text{E}][\text{S}][\text{A}]}{K_\text{m}\alpha K_\text{x}}}{[\text{E}] + \frac{[\text{E}][\text{A}]}{K_\text{x}} + \frac{[\text{E}][\text{S}]}{K_\text{m}} + \frac{[\text{E}][\text{S}][\text{A}]}{K_\text{m}\alpha K_\text{x}}}$$

Divide top and bottom by $\frac{[\text{E}]}{K_\text{m}}$

$$\frac{v}{[\text{E}]_\text{t}} = \frac{k_\text{cat}[\text{S}] + \beta k_\text{cat}\frac{[\text{S}][\text{A}]}{\alpha K_\text{x}}}{K_\text{m} + \frac{K_\text{m}[\text{A}]}{K_\text{x}} + [\text{S}] + \frac{[\text{S}][\text{A}]}{\alpha K_\text{x}}}$$

$$v = \frac{V_\text{max}[\text{S}]\left(1 + \frac{\beta[\text{A}]}{\alpha K_\text{x}}\right)}{K_\text{m}\left(1 + \frac{[\text{A}]}{K_\text{x}}\right) + [\text{S}]\left(1 + \frac{[\text{A}]}{\alpha K_\text{x}}\right)}$$

Which is Equation 7.2.

Appendix G

Useful Software Packages for Analyzing Enzyme Kinetic Data

G.1 Desktop Based Tools

GraphPad Prism

GraFit

Laboratory Guide to Enzymology, First Edition. Geoffrey A. Holdgate, Antonia Turberville, and Alice Lanne.
© 2024 John Wiley & Sons, Inc. Published 2024 by John Wiley & Sons, Inc.

Kintek Explorer

Dynafit

G.2 Web-Based Tools

Enzo

ICEKAT

G.3 Other Useful Online Resources

OmniCalculator

BioNumbers

NIST Reference on Constants, Units, and Uncertainty

6.3 Other Useful Online Resources

OmniCalculator

Bloomberg

Index

Note: *Italic* page numbers refer to *figure* and **Bold** page numbers reference to **tables**.

Laboratory Guide to Enzymology, First Edition. Geoffrey A. Holdgate, Antonia Turberville, and Alice Lanne.
© 2024 John Wiley & Sons, Inc. Published 2024 by John Wiley & Sons, Inc.

Printed and bound by CPI Group (UK) Ltd, Croydon, CR0 4YY

27/10/2024

14580681-0001